Konstantin Eduardovich Tsiolkovsky

Konstantin Eduardovich Tsiolkovsky

The Pioneering Rocket Scientist and His Cosmic Philosophy

Daniel H. Shubin
Editor and Translator

Algora Publishing
New York

Library of Congress Cataloging-in-Publication Data —

Names: Shubin, Daniel H.
Title: Konstantin Eduardovich Tsiolkovsky : the pioneering rocket scientist and his cosmic philosophy / Daniel H. Shubin, editor and translator.
Description: New York : Algora Publishing, [2016] | Includes bibliographical references.
Identifiers: LCCN 2016036590 (print) | LCCN 2016038382 (ebook) | ISBN 9781628942378 (soft cover : alk. paper) | ISBN 9781628942385 (hard cover : alk. paper) | ISBN 9781628942392 (pdf)
Subjects: LCSH: TSiolkovskii, K. (Konstantin), 1857-1935. | Aerospace engineers—Soviet Union—Biography. | Astronautics—Russia—History.
Classification: LCC TL789.85.T8 K59 2016 (print) | LCC TL789.85.T8 (ebook) | DDC 629.4092 [B] —dc23
LC record available at https://lccn.loc.gov/2016036590

Printed in the United States

Other books by Daniel H. Shubin

A History of Russian Christianity (in 4 volumes)
- Vol. I. From the Earliest Years through Tsar Ivan IV
- Vol. II. The Patriarchal Era through Tsar Peter the Great: 1586 to 1725
- Vol. III. The Synodal Era and the Sectarians: 1725 to 1894
- Vol. IV. The Orthodox Church, 1894 to 1990: Tsar Nicholas II to Gorbachev's Edict

Daniel and Alla Andreev

Leo Tolstoy and the Kingdom of God within You

Skovoroda: The World Tried to Catch Me But Could Not

Helena Roerich: Living Ethics and the Teaching for a New Epoch

Russia's Wisdom

Tsars and Imposters: Russia's Time of Troubles

Aleksandr Dobrolubov—Russia's Mystic Pilgrim

Monastery Prisons

The Gospel of the Prince of Peace, A Treatise on Christian Pacifism

Concordia Antarova

Rose of the World (Tr.)

The Grey Man, Menace Eastern-Light (Tr.)

Table of Contents

Photograph taken 1924.

2. In his library

3. With His Hearing Aid

4. With Early Designs of his Dirigibles

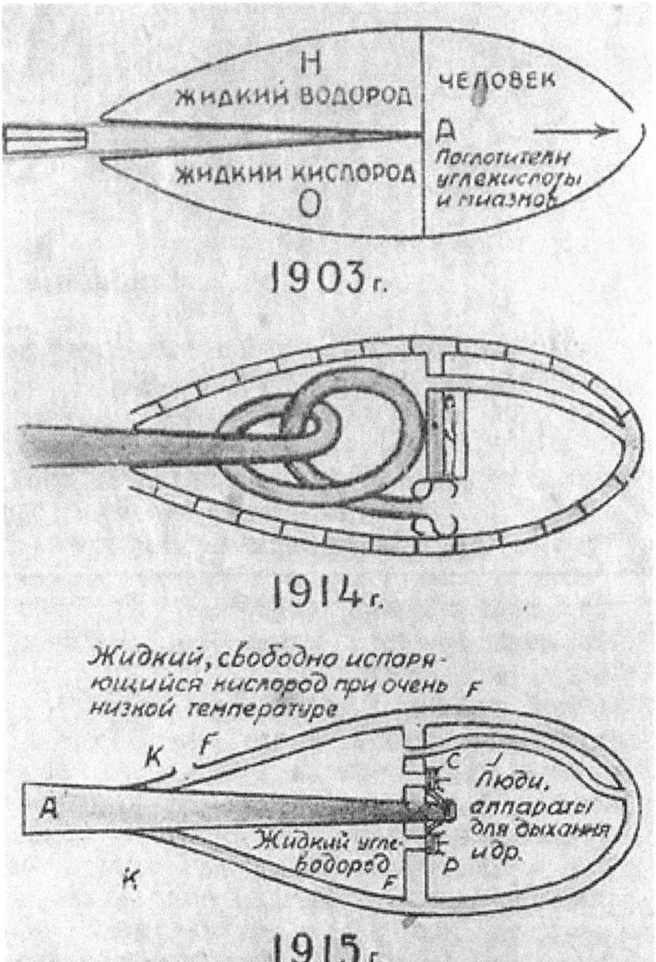

5. Early Designs of Jet Propelled Rockets

6. Konstantin and Varvara Tsiolkovsky with their four sons and three daughters.

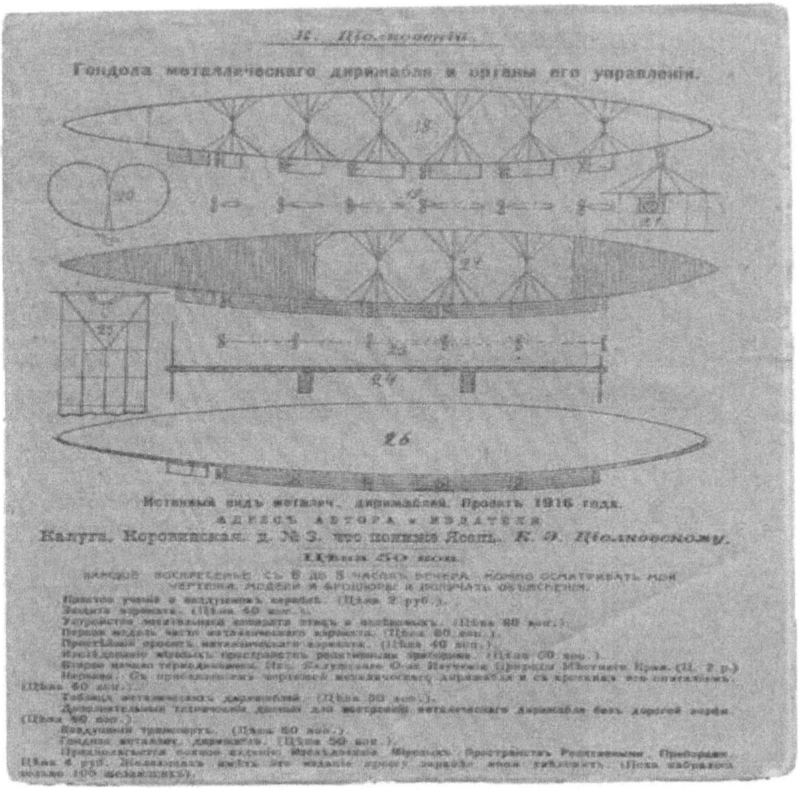

7. Over the years, Tsiolkovsky built and tested about 100 model aerostats of various shapes and sizes and lengths to determine the best design. Some of them are depicted in the above page in one of his brochures.

Introduction

The founder of Soviet astrophysics and cosmonautics was a self-taught scientist, inventor, schoolteacher, philosopher and science fiction writer. This book is a portrait of an extraordinary man and his extraordinary mind. It provides a sketch of his life and an introduction to his views and ideology, and a collection of original translations of his scientific work, philosophy, and science fiction stories.

Who was Tsiolkovsky

Konstantin Eduardovich Tsiolkovsky was a pioneer in astrophysics and cosmonautics. His pioneering creations included important designs: dirigibles formed with a metallic shell and a unique navigation system, the jet propulsion engine, the use of rockets for space travel, high altitude balloons, air-cushion vehicles, high-speed trains, and heavier-than-air aircraft.

His scientific studies contributed to the advancement in Soviet Russia of cosmology, aeronautics and aerodynamics, space exploration, atomic research, jet fuel, and the manufacture of rockets and dirigibles. Later in life he motivated the younger generation of Soviet readers with his science fiction stories of travel to the moon and throughout the solar system.

Unique to Tsiolkovsky was his conviction that advanced life existed on other planets and in other planetary systems throughout the Milky Way and in other nebulae, and his belief in the ability of humans to progress toward the settlement and development of other planets with the creation of advanced civilizations.

Tsiolkovsky was a confirmed socialist and after the Revolution he utilized his scientific studies for the advancement of socialism in Russia. As with many scholars and scientists of his era, Tsiolkovsky was against war, violence and oppression. Tsiolkovsky was also vegetarian. He considered even the killing of animals repulsive.

His later treatises advocated freedom of expression and research, and he promoted the rights of women and minorities. He believed it was a requirement of the state to assist the underprivileged and provide all basic needs to the entire population, including education, housing and security.

Tsiolkovsky's Ideology

The passages selected for inclusion in this volume provide an excellent survey of Tsiolkovsky's ideology and only a brief review will be presented here.

The universe is all matter and it has existed infinitely into the past and it will exist infinitely into the future. The size of the universe is infinite, immeasurable and unending. This universe consists of planets with their satellites, suns and solar systems and the other objects of outer space, nebulas and galaxies, and groups of galaxies which Tsiolkovsky calls astronomical units.[1] The number of astronomical units is likewise immense and immeasurable. Tsiolkovsky calls groups of astronomical units—Ethereal Islands, and their sum is the entirety of the universe.

With a galaxy containing billions of solar systems, and an astronomical unit containing billions of galaxies, the Ethereal Island contains millions of astronomical units. The universe likewise contains an immense quantity of Ethereal Islands. This is the size of the universe.

Time is just as infinite as the universe, without a beginning and without an end.

What drives the universe is happiness, otherwise known as the fundamental law of the universe. This is the essential intangible force which causes the universe to remain in motion, as it has done infinitely into the past and will so do infinitely into the future. This intangible force will never vanish or fail but will drive the universe forever.

Apart from the intangible force of happiness, the tangible force that drives the universe is gravity, which is the source of all energy in the universe. It is a combination of both these forces that life is periodically generated, along with suns and planets and all other objects of interstellar space. Every atom contains a gravitational force and this is the reason for the creation of elements and compounds and all matter in the universe. Atoms, which are matter, are the composition of the corporeal universe.

[1] This is not to be confused with the definition applied by other scientists as the distance between Earth and Sun. Tsiolkovsky has his personal definition.

Heat is energy and it is always conserved, it is never lost; and heat, because it is attached to an atom, likewise becomes a tangible item subject to the force of gravity. Heat that appears to dissipate is collected by atoms and conserved and reutilized.

All the universe operates in a cyclic pattern, all matter iterates on a periodical basis. Atoms are not destroyed but recycled. Atoms combine and separate over the interval of time, and time is not thought of in terms of hours or years or even millennia, but in large exponential numbers, such as decillions of decillions of years.[1] To understand the course of the universe, time in these terms needs to be utilized.

Where did matter originate? Matter in the form of heat-retaining atoms always existed and they collected from regions of the universe, and due to the gravitational pull between them they conglomerated and compressed, in the process forming the various elements, causing them to connect to one another.

The present solar system evolved from a concentration of matter and heat due to an extreme amount of gravitational force. Due to extreme gravitational force and extreme heat concentrated within this mass of extremely high density, the mass then reached a point of super-compression and then ignited and subsequently exploded. So did the sun ignite, and the planets and other spatial objects were created as a result. They subsequently cooled during their projection and went into all their respective orbits.

The same applies to a galaxy, but on a larger scale.

When the heat source dissipates, the inherent gravitational force of the objects will again repeat the above scenario. The iteration causes elements and compounds to form.

Each atom has the ability to maintain sensation, it is animate. An atom is not just a particle of matter, it is alive. Life evolves when sufficient atoms merge together into molecules and the molecules merge together into a complex inorganic substance, but then under the proper conditions on the planet that is conducive to life, the atoms with its life-generating ability evolve into a simple organic entity. Over prolonged time the organic entities develop further and become more complex organisms: plants and lower forms of animals and then the higher forms of animals.

Life periodically generates since the fundamental law of the universe is happiness. Every planet having the proper location in its respective solar system with the proper environment will generate life and life will progress from its basic single-cell form to the perfection of a human being, and will continue to self-perfect as time progresses into advanced entities.

Happiness causes this to iterate on an infinitely periodic frequency.

[1] A decillion is 10 to the 33rd power (1 with 33 zeros behind it).

Death is purely a mechanical function. It is a dissolution of the elements of the living object. But life does not disappear, because every atom contains life, and since it was the right combination of atoms and elements and compounds in the right environment that generated life, it recycles in the universe to eventually generate new life. This has occurred from time infinitely in the past and will to time never ending.

Happiness as the fundamental law of the universe causes life to periodically generate, to populate the planet, to reproduce plant life and animal life, and this will continue until the heat source stops, and then the cycle will start again. Such intervals though occur over immense lengths of time defined by decillions of years.

There is no such thing as spirit as defined by religion. What does exist is very fine or tenuous gas. Supernatural beings likewise do not exist, however it is possible to have live entities composed of such low density or tenuous matter that they appear to be invisible. As far as deity is concerned, a personal God as defined by religion does not exist. The determining regulation is the cause of the universe, as described in Tsiolkovsky's compositions. However, entities can exist in the universe and have an effect on our lives, but these are entities that have progressed their development over the decillions of years to the point that they have become semi-deified and whose composition is low density matter.

The Selection of Works Translated in this Volume

I have selected a cross-section of articles that explain Tsiolkovsky's philosophy and technology in the clearest possible manner, without being redundant or verbose. As this information can be acquired from other sources my biography of Tsiolkovsky will not cover the technical aspects of his theories or calculations, , although a few will be generalized.

Tsiolkovsky in his original compositions utilized the units of measure that were in use during Imperial Russian, and later he utilized the decimal system introduced into Soviet Russia in 1918. For the present work, all of these have been converted to the units of measure standard in the United States, unless indicated in the footnotes.

After reading the Introduction, I recommend the reader turn to Tsiolkovsky's three autobiographical compositions; then proceed to the biography; then the balance of the selections. Tsiolkovsky's science fiction is best read last. This will give the reader the clearest understanding of the man, his life, his concepts and discoveries, and his futurity, in a progressive manner.

All of the texts were translated from the original Russian by the present author-translator. Any additions to the text for clarification are stated in [brackets].

Only original Russian sources have been used in writing the biography.

Daniel H. Shubin
August 20, 2016

Biography of Konstantin Eduardovich Tsiolkovsky

Cossack heroes and Polish nobles figure in the background of the Tsiolkovsky family. By tradition they trace their genealogy to Cossack peasants living in a region of the Polish-Lithuanian Commonwealth that is today known as northwestern Ukraine, a region that was part of the Russian Empire in earlier periods. They claim the infamous Cossack hero Severyn Nalyvaiko as their progenitor.

As Hetman of the Ukrainian Cossacks, Nalyvaiko (or Nalivaiko) was a leader of the anti-feudal peasant rebellion during the years 1594–1596. His descendants were exiled to the Płock Voivodeship in Poland, living in the village of Ciołkowo (in Polish), or Tseolkovo (in Russian)—the origin of the family name. There they advanced to the status of Polish nobility. This account has been preserved as family legend and is not verifiable by historical research.

Based on historical accounts of the era, the Ciołkowski family evolved from Polish nobility that migrated to Russia in earlier centuries. The name was then rendered in the Cyrillic alphabet, and from that time on the common transliteration is Tsiolkovsky.

The earliest mention of the Ciołkowski name in history dates to the year 1697; members of the family—three brothers named Stanisław, Jakub and Valerian—were involved in the election of King Augustus II the Strong that year. In 1777, Jakub Ciołkowski, the great-grandfather of our subject, sold his estate and migrated to Berdichev County, Kiev province, and then subsequently migrated to Zhitomir County, Volyn Province, at the western edge of the Russian Empire. In time, the family discontinued its affiliation with its roots in the Polish nobility and assimilated into Russian society and culture.

8. Map of Russia with locations of Tsiolkovsky's residencies over his lifetime (Kirov is the Soviet-era name for Vyatka)

The father of our subject, Eduard Ignatievich (1820–1881), was also known as Makary Edward Erazm Ignatyewicz Ciołkowski on the Polish side. He was born in the village Korostyanin, Rovnensk Province in northwestern Ukraine. In 1841, he completed the Forest and Surveying Institute in Petersburg, and then served in the forestry service in the Olonetz and Petersburg Provinces. In 1843, he was transferred to the forestry service at the Pronsk Forest Reserve in Spassky County, Ryazan Province, southwest of Moscow. For the most part, Eduard was atheist. While living in the village Izhevskoye, he met his future wife, the mother of our subject, Maria Ivanovna Yumasheva (1832–1870).

Her roots were among the Mongols who invaded Russia several centuries earlier and who intermarried with the native Russians and then, during the reign of Tsar Ivan IV Vasilyevich, settled to live in Pskov province. Maria was raised in the traditional Russian fashion. Her parents were nominal landowners and were coopers and basket makers. Her parents provided Maria the best education available for a woman at the time; she finished high school and studied Latin, mathematics and other sciences.

Immediately after their wedding in 1849, the Tsiolkovsky couple moved to the village Izhevskoye, where they resided until 1860. Their first son was Dmitry, also called Mitr or Mitya. Then came a second son, Ignaty. Konstantin was their third son, born September 5, 1857, in the same village, and then another son (name unknown), and later two daughters, Ekaterina and Masha. (Tsiolkovsky in his autobiography states his mother had 13 children. This could be a typographical error, since other sources only mention six, or some may have died shortly after childbirth.)

At the age of ten, Tsiolkovsky was riding on a sled during an early winter freeze. He caught a cold and it developed into scarlet fever. After a prolonged and critical illness, Tsiolkovsky lost 90% of his hearing. This began what Tsiolkovsky called the most depressive and miserable period of his life. Partial deafness or bradyacuasia deprived the boy of most childhood games and the impressions that were available to all of his healthy contemporaries. Maria Ivanovna, his mother, spent dedicated and concentrated time to teach him to read and write, and taught him the fundamentals of arithmetic. She did all she could so he would never sense that he was handicapped.

Recollecting his father, Tsiolkovsky explains his hostile attitude toward the imperial government.

> When my father's Polish and liberal friends gathered at our house, they were decent people of the higher government and official levels. My father was never incarcerated, but he did often have altercations with the local police as well as several unpleasant disputes with state officials. This strife eventually caused his termination from the forestry service.

The Tsiolkovsky family had spent a year in Ryazan and the father, Eduard, was promoted to teach natural history at a high school there. But eventually this job failed. In 1868, Eduard and the family had relocated to Vyatka[1] to work there at the forestry service.

Tsiolkovsky enjoyed living in Vyatka, and what especially attracted him was the Vyatka River, which never froze over but had flowing water the year round; steamships navigated the river. Tsiolkovsky was already swimming the first summer after the family relocated to Vyatka. In fact, during his entire life, from this time onward, Tsiolkovsky lived near a river or a large body of water; but this also caused him grief over the years.

During the winter it was typical to go ice skiing and sledding, and in spring the boys would jump from ice floe to ice floe and hope not to fall into the icy cold water. Tsiolkovsky almost died as a result once, as he recorded in his autobiography. Tsiolkovsky also relates how he would clamber to the top of the bell tower of the ancient Orthodox church in the city center and climb on the bell, and then venture even higher to the top of the steeple. The fear of possibly falling to his death finally occurred to him, and he never did it again. Tsiolkovsky in his memoirs related having nightmares of the incident in later years.

[1] Today known as Kirov, northeast of Moscow. The Vyatka River flows through the center of the city.

Tsiolkovsky's mother died in 1870 when he was 13 years of age, just before he resigned from high school; she was 38 and had been married 21 years. A happy woman, filled with life's joys, humorous and laughing, as Tsiolkovsky described her in his memoirs, she dearly loved her son, doing all she could so he would never feel he was handicapped or underprivileged. Due to the poor relationship Tsiolkovsky had with his father, his mother's death had a devastating effect on him. Deprived of maternal support for his education, his learning ability decreased.

Even with his son's bradyacuasia, father Eduard was able to get him and younger son Ignaty enrolled at the all-male Vyatka high school, so at age 12, Tsiolkovsky was on his way to a good education. However, matters did not go well later with his inability to hear the instructor and so complete the lessons, and often he would be in the chancellor's office. Tsiolkovsky managed to complete the entirety of his first and second years, but during his third year he quit. His mother's death contributed to his decline in learning. Tsiolkovsky stated in his memoirs:

> I could not learn anything in school. Either I heard completely nothing or else just heard indecipherable sounds. But gradually my intellect found another source of information—books.

This statement also testifies that Tsiolkovsky did not have any formal education past about the age of 14, and just a rudimentary education up to that time. As a result, he always considered himself self-taught.

Tsiolkovsky related the following about his father:

> He was always cold, introverted. But he never touched or insulted anybody, but everybody was cautious when near him. We were afraid of him, although he never lost control over his temper, never swore and never fought.

The entire situation caused Tsiolkovsky to become withdrawn, offended and irresponsible, and now his hearing loss had a more deleterious effect on him, and especially realizing that he was now orphaned. However Tsiolkovsky's natural character did not allow him to capitulate to failure and depression, and he recovered from his melancholy state. Tsiolkovsky stated in his memoirs that suddenly there manifested in him the desire "to achieve great feats and earn people's approval and not to descend into a state of regret."

As opposed to the teachers at the high school, books generously shared their knowledge with Tsiolkovsky and never made even the smallest objection or criticism. Now studying on his own, he was able to learn what he could not in the classroom. The topics became clear and understandable, as Tsiolkovsky related in his memoirs:

> When I was 14 to 15 years of age, I became interested in physics, chemistry, mechanics, astronomy, mathematics, and more. Books of course were few and difficult to acquire, and so I often had to rely more on my own intuition and deliberate mentally on concepts. I progressed further than what I read, not stopping at just the context. I did not understand many items and there was no one to explain them to me, and even then, with my handicap they could not help me anyway. This forced me to develop the ability to think independently, to resolve technical questions on my own.

Thus did our Tsiolkovsky find his place in life, even with his auditory handicap: self-education through technical books. As Tsiolkovsky relates in his memoirs, his favorite text was a translation from the French into Russian of *A Complete Course in Physics*, by Adolf Gano,[1] published in Russia in 1866. The publisher was F. Pavlenkov, and he was living in exile in Vyatka at about the same time as Tsiolkovsky, and so he was able to acquire a copy from him.

It was this book that introduced Tsiolkovsky to the aerostat, as one chapter was dedicated to this vehicle. Thoroughly studying the concept and design of such a vehicle, Tsiolkovsky concluded:

> I need to note that the true use of aerostats will only occur when a means is developed to navigate them. All other attempts to use them purposefully without the ability to steer them are worthless.

So began Tsiolkovsky's lifelong dedication to the development of a navigable as well as metal-clad aerostat. His first experiment was the attempt of a small aerostat with a shell made of paper, but nothing came of it since hydrogen was not available for the general consumer. He also realized that it would not have flown anyway since ordinary paper was not strong enough to retain the pressurized hydrogen. This led Tsiolkovsky to the conclusion that any steerable aerostat would need to have a metal shell.

While studying one of his father's textbooks on the topic of land measurement, which he acquired while working for the forestry service, Tsiolkovsky became interested in determining the distance between two inaccessible objects. Based on drawings and descriptions in the book, he was able to devise an angular instrument—an astrolabe. Of course, his homemade astrolabe was a far cry from the instruments used by surveyors. But nonetheless it was functional and he utilized it to determine the distance to a lighthouse, which he calculated to be about 400 yards. Then he measured the distance by walking and the two distances concurred.

Tsiolkovsky's habit of doing everything with his own hands materialized shortly after he lost his hearing. As an adolescent he enjoyed building his own toys, and paper and cardboard served as his materials, cut precisely using

[1] French physicist and meteorologist (1804–1887).

scissors, while sealing wax and glue connected the parts. From his capable hands small houses, sleds, and even clocks with gears were assembled. Tsiolkovsky purchased used crinolines at the bazaar at a low price and disassembled them and used the fabric for his hobby and inventions. He wound and made springs out of the metal hoops, which he used in his self-propelled—meaning horseless—carriages and locomotives. The most important single tool in his workshop was his turning lathe.

When father Eduard saw his son's workshop, his attitude completely changed toward his son, now realizing his capacity for invention and innovation and ability to study and work independently. Eduard made the decision to send his son to school in either Moscow or Petersburg. He was hoping his son would enroll at the Higher Technical Institute (today known as the Moscow State Technical Institute). After discussing the matter, the father was able get some letters of recommendation for an easy entrance into the institute and Tsiolkovsky departed Vyatka for Moscow with the father contributing expense money every month. He was 16 years of age and the year was 1873.

According to Tsiolkovsky's memoirs, Moscow was not a welcoming place. The arrogant denizens distrustfully stared at the poorly dressed provincial, while the rest just ignored him. Taxi drivers looked at him and continued, knowing by the way he was dressed that he could not afford them.

Somehow he lost the letters of recommendation and once arriving at the school, the place intimidated him and he never entered the facility. Eventually he rented a room in the home of some elderly woman. Initially Tsiolkovsky spent most of his time at the Chertkov Library, which closed in 1874, and all of its contents were moved into the existing and nearby Rumyantzev Library, where the famous futurist Nikolai Feodorovich Feodorov, the Muscovite Socrates, was librarian. Other famous figures also patronized the library, such as Leo Tolstoy, Aleksandr Ostrozhsky, Ivan M. Sechenov, Dmitry Timiryazev, Nikolai Yegorovich Zhukovsky and Aleksandr Stoletov.

Tsiolkovsky's working day began early. The distance from his room to the library was considerable, and he could not afford a horse-drawn carriage. At 10 AM, when the library doors were to open, he was standing ready to enter. He spent the entire day studying at the library until closing time at 4 PM.

While frequenting the Rumyantsev Library he met and acquired an association with Nikolai Feodorov and had many discussions with him. Feodorov recognized scientific talent in Tsiolkovsky and so allowed him access to many books on scientific subjects that were not regularly supplied to other patrons. Of course, Tsiolkovsky was quite appreciative that someone whom he considered advanced in the spheres of futurist science would condescend to encourage him by supplying him with technical books.

In his first year in Moscow libraries, studying on his own, the diligent Tsiolkovsky grasped concepts of physics and mathematics. During his second year he studied differential and integral equations, higher algebra, analytic geometry, and spherical trigonometry, all from books, and mastered them all.

The second half of Tsiolkovsky's day, after the library closed, was returning to his room and experimenting with all of the ideas that materialized as a result of his studies, as well as theories and calculations to justify them, most of which dealt with aerodynamics.

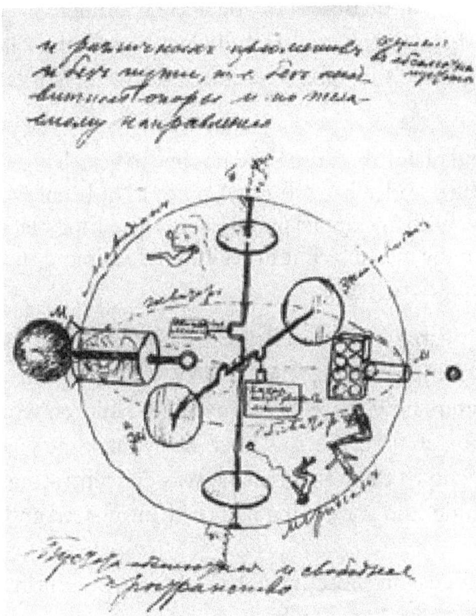

One of his deliberations dealt with the possible use of centrifugal force to rise above the atmosphere and into outer space. The sketch above is Tsiolkovsky's initial design of rotating discs attached to the end of revolving pendulums, and all of this attached inside a spherical shell. Somehow the fast velocity of the discs and the rotating pendulums, causing a massive centrifugal force, would cause the device to rise against gravitational force and propel the device into outer space.

Ecstasy overwhelmed Tsiolkovsky when he completed this design of a spaceship. Regardless that it was late evening when he finished it, Tsiolkovsky could not remain in one place, so the young man left his room and wandered about Moscow at night, fantasizing himself a cosmic traveler. However after about an hour of wandering the city streets, he realized the mistake in his design: that something rotating inside a shell is not going anywhere on its own. Tsiolkovsky recorded in his memoirs:

> However short the ecstasy, it made so strong an impression on me that I would see this device in dreams for the rest of my life. I was mesmerized by it and actually imagined myself inside of it rising high.

Such is an example of the type of innovative intellect that Tsiolkovsky possessed and which continued to motivate him further with other ideas, some just as ludicrous, but some were prophetic or intriguing, and some materialized with a practical application. But it all started with his fantasies.

Father Eduard sent his son 10 to 15 rubles a month. Tsiolkovsky spent the money on technical books, chemicals, laboratory appurtenances, instruments and equipment. He survived literally on bread and water. Every three days he would buy a bread roll for 9 kopeks, so one ruble covered 10 rolls for the month's food. The balance he spent on his education and experiments.

After several years passed, Tsiolkovsky cheerfully laughed at himself, at what he had endured during these years for the sake of science. He related in his memoirs:

> My hair was long, and plainly because I had no time to get a haircut. I walked about disheveled and what a fright I must have been. But nonetheless I was happy with my fantasies and ideas, and eating black bread did not seem to annoy me. In fact, it never entered my mind that I was hungry and was emaciating myself.

Hungry, in a tattered suit eaten away by chemicals, covered with sores from reactions of his skin with chemicals, Tsiolkovsky walked about the city. He recorded in his memoirs how shameless Muscovite children would follow behind him and shout at him, "So did mice eat your pants?" But Tsiolkovsky did not condescend to their antics as he was concentrating on his experiments for that evening, and something had to be purchased and he needed to stop at a bazaar.

During Tsiolkovsky's residency in Moscow, Nikolai Feodorov continued to provide him with books on science and technology otherwise prohibited from distribution; this further developed Tsiolkovsky's ambitions for invention and discovery. At the same type Tsiolkovsky devoured cultural literature, such as Leo Tolstoy, Turgenev and much Shakespeare, and also started to read revolutionary literature, and especially Dmitry Ivanovich Pisarev. Eventually Tsiolkovsky would become a dedicated communist and supporter of the revolution and the new Soviet government. He recorded in his memoirs:

> The populist publicist Pisarev forced me to shake from joy and happiness. In him I saw myself, my second *me*. He was one of the most respected of all of my political commentators.

Father Eduard of course was expecting his son to enter one of the universities in Moscow to further his education, but this was not to happen. Tsiolkovsky felt his independent instruction provided him more than any classroom, especially since his auditory handicap meant he would not be able to learn anyway and most likely would be expelled, as he had been in Vyatka. His new attraction to revolutionary literature also worried father Eduard, and based on correspondence between father and son, Eduard decided to stop subsidizing his son in Moscow and force him to return. In any case the elderly father was hardly able to subsidize his son, having his own health problems and no work. The difficult Moscow years, the long hours in poorly illuminated libraries and poor living conditions, caused Tsiolkovsky to lose some of his vision, and from this time on he was forced to wear eyeglasses.

Although not wanting to leave Moscow, Tsiolkovsky had no choice under the circumstances. When he returned to Vyatka, he was pale and underweight. It was September or October 1876, and Tsiolkovsky was 19 years of age. It was time for him to get a job and support himself.

As a result of his education, Tsiolkovsky offered himself a tutor for failing students, and he took whatever they paid him for his mentoring. Even then, any spare or idle time was spent in the local Vyatka library. According to his memoirs he consumed Isaac Newton and the Russian mathematician Nikolai Brashman, as well as reading the local progressive magazines on political issues, including Konstantin Stankiuvich, Dmitry Pisarev and Alexandr Sheller. His most valued textbook was Newton's *Mathematical Principles of Natural Philosophy*.

He mentions in his memoirs the practical results of his studies:

> These magazines had an immense influence on me. For example, reading an article against tobacco, I decided not to smoke. This decision I held my entire life.

With the money he earned from tutoring, Tsiolkovsky rented a room and moved all his equipment, tools, instruments and books into it and turned it into a workshop. His first attempt at an original invention were water skis with a person sitting in a chair and peddling and operating gears that turned a fan for propulsion. But it failed.

Tsiolkovsky's brother Dmitry died in 1869, and his sister Ekaterina died in 1875. His brother Ignaty died in November 1876 when typhus plagued the region. Tsiolkovsky was now left without a mother or siblings. The several deaths in the family left his father Eduard devastated. In 1878, he decided to move to Ryazan, back to his previous residence, and to quietly spend the balance of his life there. It was a mistake, according to Tsiolkovsky's memoirs; he was disappointed, and the move made life more difficult for his father. Shortly after arriving in Ryazan, Tsiolkovsky moved away from his father and rented a room, again setting up his workshop and study there. He started his independent study to pass an examination, hoping to become a teacher in a local county school. Since the Tsiolkovsky family were not practicing Orthodox Christians—Eduard was a borderline atheist—the most difficult part of the examination was Catechism, but he passed successfully and received credentials to teach arithmetic and geometry.

After moving to Ryazan, Tsiolkovsky sketched a chart of the solar system with a drawing of the orbits of all the planets. He listed all the objects in the planetary system, such as satellites and asteroids with their location and orbit. One purpose of his scheme was developing a means of hopping from one cosmic object to another during space travel. He also listed the surface area, volume and mass of all the planetary objects, their density compared to water, their relative gravitational force compared to Earth's, and the attraction of the planet to its satellites and any attraction the planet may have on objects others then its satellites. In his chart were the planets' solar cycles, the rotation of the satellite around its planet, the planets' and satellites' individual axial rotation and speed of rotation on its surface at its equator, and the orbit of the satellites relative to its mother planet.

The chart was immense when completed and it testified to the serious nature of Tsiolkovsky in his quest of discovery and advancement of space travel. He was 20 years of age at this time. Tsiolkovsky mentions the following in his memoirs:

> Astronomy attracted me and enthused me, because I considered and still consider to this time that not only Earth, but the entire universe is capable of human residency.

Still waiting for an assignment as a mathematics teacher, Tsiolkovsky did not allow time to pass uselessly and he continued to apply his theories to experimentation. He constructed a centrifuge for humans to experience force from acceleration and high velocity. (The same design was used later by Soviet scientists while training cosmonauts.) Tsiolkovsky later recollected what he had done:

> For a long while I did experiments with different animals, subjecting them to the effect of a compressive force on special, centrifugal machines. What was good is that I was not successful in killing even one live animal, even though this was not my intent to go to the extent of one actually dying in the experiment, but I did think this might occur. I remember a red cockroach that I found in our kitchen, that I subjected to an increase of 300 times the force of gravity, and then a chicken subjected to 10 times the force. I did not notice any harmful effect on them caused by the experiment.

In Tsiolkovsky's eyes, such an experiment was a tremendous advancement in aerodynamics, but it would evade any mention in the newspapers and scientific journals. In 1879, the fate of cockroaches and chickens subject to an increase in gravitational force interested none, or few if any.

It was not until January 1880 that Tsiolkovsky received his long-awaited assignment from the Ministry of Education, and it was in Borovsk, but not too distant from Ryazan. He only lived in Ryazan less than two years before he moved, and Tsiolkovsky's assignment was to teach arithmetic and geometry at a county school in Borovsk.

It was this year that Tsiolkovsky composed his first scientific article, titled *Graphic Portrayal of Feelings*. This first article was pessimistic and reflected the recent tragedies of his family, applying the philosophy of fatalism to the progress of universal history. A person is subject to forces beyond his control and life can seem meaningless; however, since the universe is timeless and infinite, the goal is to persevere and proceed. Eventually the person will merge into the state of the universe. This philosophy would accompany Tsiolkovsky through the end of his life and encourage him and allow him to persevere.

His father Eduard died January 1881 in Ryazan. This left Tsiolkovsky entirely alone.

Borovsk was a city populated primarily with Old Believers or Old Ritualists (Staro-Veri), those who observed Russian Orthodoxy using the rites and traditions before the renovations of Patriarch Nikon about 200 years earlier. Many were apprehensive of Tsiolkovsky, who was borderline atheist. Nonetheless he was able to rent a room at the house of a widowed Old Believer priest who was part of the Unity-Faith movement (Edino-Veria), an effort to allow the old rites and traditions but still come under the jurisdiction of the Patriarchal Orthodox Church. This was Evgraf Nikolaevich Sokolov, who had one daughter, Varvara (sometimes called Varya); she was two months younger than Tsiolkovsky. He found himself quite comfortable there and was attracted to Varvara's diligence and pleasing character, and as usual he was not far from a river, in this case, the Protva River. Eventually and inevitably, Tsiolkovsky concluded Varvara would be a good match for him.

Tsiolkovsky related his time at their home and the relationship in his later memoirs:

> I conversed with him lots over tea, lunch or dinner and with his daughter. I was struck with amazement at his understanding of the Gospels.

> It was time for me to marry and I married her without love. I was hoping that such a wife would not interfere with my life's calling, but would work on my behalf. My hope was fully realized. We walked about 3 miles to the church to marry, on foot, not especially dressed for the occasion, and we did not allow anyone else into the church. We returned and no one knew anything about us getting married.
>
> I never knew any woman in my life, before marriage or after, other than my wife.

Konstantin and Varvara were married by her father at a local Orthodox Church on August 20, 1880. It was a private wedding with no reception following. They would be married 55 years until his death and would have seven children: four boys and three girls. In Borovsk their initial four children were born: daughter Lubov (1881) and sons Ignaty (1883), Aleksandr (1885), and Ivan (1888). One son, Leonty (1892) and two daughters, Maria (1894) and another, about whom little is known, were born in Kaluga.

Konstantin and Varvara had a tragic and difficult time with their family. Of their seven children, all four sons and one daughter would eventually die during their parents' lifetime. Son Ignaty (in 1902) and son Aleksandr (in 1923) both committed suicide. Leonty died from whooping cough on his first birthday, and Ivan died in 1919 from malnutrition during the Russian Civil War. One of the younger daughters died from natural causes.

Typical of Tsiolkovsky, while returning home from the wedding he stopped at a store and purchased a turning lathe, another one, for manufacturing parts for his experiments. He had left the first one behind in Ryazan. Married life did not deter Tsiolkovsky as he continued to use funds available—after basic family and household needs were met—for his experiments and projects. He had lived the previous 10 years—since his mother passed away—for the most part as an isolationist and independent citizen, and now continued much the same course. Several years later, Varvara Evgrafovna recollected her wedding day in the following manner:

> We did not have a wedding reception, and he even forgot to take the dowry with us to church.[1] Konstantin Eduardovich said that this is how we would be living: modestly and as far as his paycheck would go.

The new couple rented a room near the school where Tsiolkovsky taught.

Daughter Lubov Konstantinovna wrote about her father's attitude toward religion:

[1] In Russian tradition, the husband provides a dowry for the wife.

> He had a negative attitude toward the superficial ceremonial side of religion. He only attended church when officials would remind him about it. But recognizing a person's freedom of opinion, including religion, he never interfered with my mother attending church herself, or any of us, his children, should we decide to attend also. He considered it our entertainment. But he did consider churches to be a nice decoration of cities and as ancient monuments. He listened to church bells as though it was pure music, and he enjoyed wandering about the city during vespers.
>
> He considered Christ a great humanitarian and an ingenious individual who was able to instinctively foresee truth. So for example he saw in the statement of Christ, "In my Father's house are many residences," the idea of a large number of residences in the universes.
>
> In regard to morals, father placed Christ irreproachably high. His death for his ideals, his sorrow for humanity, his ability to understand all things, his ability to forgive, brought father to a state of ecstasy. But father had the same admiring attitude toward those who abnegated themselves in their dedication to science, those who delivered humanity from death and illness, toward the inventors who would ameliorate human difficulties.
>
> He believed in higher perfected entities living on planets that were more ancient than our Earth. But he thought of them as entities consisting of the same material as the entirety of the cosmos, and which according to his understanding were guided by laws, general to all the universe.

This attitude of Tsiolkovsky toward religion often caused misunderstanding among those who surrounded him. Varvara's relatives only consented to her marriage with an atheist—as they considered him—because she had no dowry to offer and Tsiolkovsky was the only person who did not consider this of any importance.

However, accidental comments that Tsiolkovsky made about Christ almost cost him his job on several occasions. At his own expense, Tsiolkovsky ended up having to travel to Kaluga while living in Borovsk to explain his comments to the satisfaction of the inquest board.

Tsiolkovsky recollected the period of his early life in Borovsk in his memoirs:

> I returned to both my toys and to serious mathematical problems.
>
> I had electric bolts of lightning flashing in my house, thunder as a result, bells would automatically ring, papers dolls I had dancing.

> Visitors were amused and entertained and were surprised at the electrical octopus I invented. I had it grab at things with its tentacles, or at someone's nose or finger, and it would sit on someone's lap, and I put hair on it and used static electricity to raise its hair, and then get sparks to fly from different parts of its body.
>
> I filled a rubber balloon with hydrogen and carefully connected a paper boat containing sand to its bottom and which balanced it. As though it was alive, it floated from room to room following the air current, rising and falling.

Such toys of a mechanical design made the arithmetic and geometry teacher popular among the residents of Borovsk.

One of Tsiolkovsky's favorite technical books was the university physics course by Professor Feodor Fomich Petrushevsky. Studying this book, Tsiolkovsky discovered the hint that led to his kinetic theory of gases, and although the professor proposed this to his readers as nothing more than a doubtful hypothesis, Tsiolkovsky was stimulated by it. It was in 1881 that Tsiolkovsky made a dedicated effort to developing his theory, much of it based on Petrushevsky.

Tsiolkovsky was convinced that he had resolved the secrets of gases and passionately considered this his first scientific theory. After he completed his treatise, titled *Theory of Gases*, a student friend of his, Vasily Vasilyevich Lavrov, took the manuscript to Petersburg to be reviewed by other professors, and it eventually made its way to the Russian Physics–Chemistry Society recently founded by Mendeleev. However the Petersburg professors saw nothing new or innovative in his discoveries; he only explained things in a different fashion than others who were already published. The reason was that, living in Borovsk, Tsiolkovsky did not have access to the books of other physicists and so took an independent approach to the development of his theory.

But the introduction of his treatise and his way of dealing with the subject caused the members of the society to take further notice of Tsiolkovsky. On October 26, 1882, Prof. P.P. Van der Flitt reported his opinion of Tsiolkovsky's research at a meeting of the physics department of the society:

> The information in his research on its own does not present to us anything new and the conclusions therein contained are not entirely precise, but nonetheless it discloses the great abilities of the author and his diligence, since the author did not receive a higher education and was able to develop and conclude such concepts based on information he acquired on his own. Certain elementary technical textbooks, such as the physics course of Prof. Petrushevsky, and the *Basics of Chemistry* of Prof. Mendeleev, served as the author's sole sources for

> the presented research. In view of this, we will be very interested in promoting further research to be undertaken by the author.

In addition to the above compliment, the society invited Tsiolkovsky to move to either the Petersburg or Moscow area, where he could more effectively utilize his scientific research capabilities. However, Tsiolkovsky never bothered to respond to the invitation, since he could not afford to move anywhere and could not afford the dues for membership in the society. Rather than admitting this, he just dropped the matter; but Tsiolkovsky still considered himself a victor, as he had gained the respect of these scientific minds. In later years, Tsiolkovsky recorded the reason for his difficulties:

> Books in general were few and difficult to acquire and for me especially. This forced me to think more independently and often I found myself going in the wrong direction, chasing some mistake. Often I would invent or discover something that was already long available or known. I learned to create, although often unsuccessfully and tardily. So I got into the habit of thinking and alluding toward matters ironically.

A storm of thoughts raged in Tsiolkovsky's head as he read the letter from the Russian Physics–Chemistry Society. Valuing the respect conveyed to him in this letter, he composed another treatise, but on a different subject, titled *Mechanics Applied to Moving Animals*. This was reprinted in 1920 under the new title of *Mechanics in Biology*. The treatise dealt with the movement of boats and fish in water relative to the resistance of water on their size and design. To over-simplify his theory, what Tsiolkovsky concluded was that long and narrow fish swim faster than short fish—and the same would apply to boats. He applied the same theory to dirigible design in later years.

The treatise was reviewed by Ivan Mikhailovich Sechenov, who considered it very interesting. Others, however, did not, and so the treatise was not published at the time. Sechenov and Tsiolkovsky began a close professional relationship at this time. Tsiolkovsky's invitation to join the society inspired him to further study, experiment, theorize, and build models of his designs.

Tsiolkovsky's third independent research treatise composed in Borovsk was *Unrestrained Space*, the word to be understood as gravity-free. It relates a fictional diary of a scientific explorer who was able to journey into outer space on a rocket ship and experience a gravity-free environment. This was also Tsiolkovsky's introduction of his concept of using jet propulsion for outer space travel, since any other method would be ineffective in a vacuum. He started this fictional diary on February 20 and ended it on April 12, 1883.

Tsiolkovsky recorded later in his memoirs:

> When I proved that outer space is not so infinitely distant and inaccessible to humanity, even though it does not seem this way at present, then such freedom of travel will earn more of the reader's serious attention and interest.

As is very apparent, even in 1883 Tsiolkovsky placed a task distantly ahead of him to resolve. Just a design for its materialization would take him years. A journey to the moon is fantasy, so Tsiolkovsky had to convey it in a literary form, and this was his science fiction book, *On the Moon.* It was written in the form of a dream that a person had when he fell into a delirium from a fever. In the dream he arrives on the moon with a fellow physicist. After traveling about the moon's surface and the unseen back side, he returns to Earth in the form of awakening from his dream when he recovers from his delirium. It would not require readers to have much depth of knowledge to understand his conception of reduced gravity on the moon, so the book was very popular. It also sparked people's interest in the moon.

Tsiolkovsky continued nonetheless to amuse the residents of Borovsk as he had those in Ryazan. The first summer he built some strange-looking boats that seemed to flow through water faster than others. The following winter he constructed a large umbrella, tied a large chair to some snow skis, and went sailing down the ice of the river.

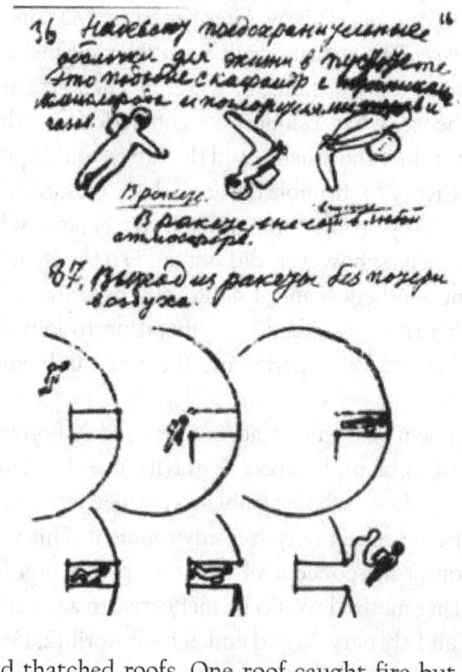

9. Tsiolkovsky's Sketch of a Door with Hatch by which to Move in and out of a Space Capsule

The one incident that got Tsiolkovsky into trouble was when he made a balloon using paper glued together, with an opening at the bottom. He placed smoldering embers in a pan under it and tied it to the balloon; the balloon rose and flew away. But then the embers burned the strings and caught the balloon on fire, and sparks fell from the sky onto people's wood and thatched roofs. One roof caught fire but it was quickly extinguished. Tsiolkovsky was arrested but the prosecution did not succeed, as everyone recognized this as a practical joke typical of him. Some in Borovsk considered him a holy fool, while others felt he was ingenious, a mad scientist, while attempting to overcome his auditory impairment. But the students at the school where he taught loved him.

It was such experiments and such fantasies that moved Tsiolkovsky toward the direction of a navigable aerostat with a metal shell, with what he played in the past, but now he dedicated himself to this concept seriously. One reason was that outer space seemed to be too far into the distant to be immediately accessible, while a dirigible was more practical and its concepts could also be applied to rocket science in the future (and some to submarine design). His dedicated effort and the construction of his models were two years in the making.

This concept materialized in Tsiolkovsky's article, *Theory and Experiments of a Aerostat Operating in a Horizontal Direction with an Elongated Shape*, in 1885–1886. He provided a completely new and original design and construction of a lighter-than-air vehicle with a thin metallic shell that was collapsible. The article included some sketches and important facets of his construction. This article established the fundamental design for all of Tsiolkovsky's future dirigibles.

His design for ascent and descent was very unique. Since hydrogen was flammable and helium was unavailable, Tsiolkovsky designed a bundle of tubes to fit inside the shell of his aerostat. Since a gasoline engine would be used to propel the vehicle, Tsiolkovsky visualized the hot exhaust of the motor flowing through the bundle of tubes inside the shell like a heat exchanger and heating the internal air before being released to the atmosphere. The heat dissipated into the volume of air in the shell would cause the air density to decrease proportionately and so cause the aerostat to rise. The ascent or descent of the aerostat would be controlled by the amount of hot exhaust allowed into the tube bundle, and the balance would be exhausted to the atmosphere.

Eventually word spread in the circle of physicists with whom Tsiolkovsky corresponded and he was invited to Moscow for a presentation of his design in April 1887. It was through the efforts of Pavel Mikhailovich Golubitzky, inventor of the Russian telephone, that Tsiolkovsky was able to make his presentation. Supporting him were other scientists mentioned earlier: Stoletov and Zhukovsky, and also Vladimir Michelson, a very prominent Russian physicist and meteorologist, as well as other prominent scientists.

Tsiolkovsky's report was titled *Of the Possibility of the Construction of a Metallic Aerostat capable of Changing Shape and to even become Flat*. The text of Tsiolkovsky's address was recorded and preserved by secretaries who attended.

> The 18th century inculcated listeners with their convictions and dreams of bird-like flying machines and aerostats, having some arbitrary ability to ascend and descend. The 19th century was the age of attempts and theories, and it is one of these that I propose to all of you, my most honorable audience.
>
> For transport by air I propose to you a metal-clad aerostat. Other than its external appearance, it has little common with existing gas-filled air-ships.[1] It is independent of temperature and atmospheric pressure, and is entirely filled with gas. The external shell will not implode or explode because it is made of thin corrugated metal.
>
> The use of such aerostats to transfer people and goods, according to my calculations, is 10 times less expensive then by railway or steamship. My proposed construction does not require expensive shipyards for manufacture, or hangars for storage. They can be preserved in the open, in a valley or any flat area, as long as it is protected from the wind by hills, buildings or trees.

Tsiolkovsky drew sketches of his designs on a board for the audience and his presentation interested them, although Tsiolkovsky considered his designs far from complete.

There are two basic designs of an aerostat that Tsiolkovsky presented. The first was fabricated of a thin corrugated metal without a frame that was almost flat when deflated. As the motor exhaust heated the internal tubes, this would cause the small amount of air to expand; any additional air would enter the aerostat from the opening in the bottom, a breather hole. In its fully expanded condition, it would appear as a regular blimp form. As the air would cool, the aerostat would deflate and shrink in size.

Tsiolkovsky's second design was a corrugated-metal shell with a fixed internal volume. The size of both aerostat designs would be based on the buoyancy of air at some higher temperature, enough to cause the aerostat to rise and maneuver in the atmosphere. The same corrugated shell design could later be used for a lighter-than-air gas, but it was difficult for Tsiolkovsky to determine how to maneuver the vehicle to descend without releasing the lighter gas and replacing it with heavier air. So he preferred his exhaust tubular heat exchanger.

It was about this time, in the spring of 1885, when Tsiolkovsky was 28 years of age, that he had an apparition that was impressed upon him for the rest of his life. He considered it divine and recorded it in his composition *Fate, Destiny, Fatalism* (translated in this volume). In another composition,

[1] Referring to the Zeppelin, which started design at about the same time, except it was cloth over a metal frame and contained balloons filled with hydrogen.

Tsiolkovsky mentions in passing that he saw a second divine apparition in May 1928, but he recorded no details.

Over the years, Tsiolkovsky built and tested about 100 model aerostats of various shapes and sizes and lengths to determine the best design. Some suggested again that he relocate to Moscow to continue his research and to have experienced technical and scholarly help more readily accessible. But Tsiolkovsky was unable to move again due to his financial difficulties. However, some also suggested financial assistance, but this proceeded no further, and nothing conclusive evolved from Tsiolkovsky's presentation.

The day after returning home to Borovsk, the Tsiolkovsky home caught fire and much was destroyed and ruined, including most of Tsiolkovsky's workshop, his library and documents. Fortunately, none of the family was harmed or any of the other residents in the building. Between the failed effort at acquiring financial subsidy, inability to move to Moscow, and the fire destroying so much of Tsiolkovsky's efforts, he went into depression and this interrupted his scientific research. Fortunately the manuscript of his next treatise, *Theory of Aerostats*, was in Moscow at the office of Prof. Zhukovsky.

Eventually, the Tsiolkovsky family found another residence and relocated, taking with them all that survived the fire. Tsiolkovsky reconstructed his workshop and library and continued his research. As always, the new residence was near a river.

Tragedy again struck. In spring 1889, the Protva River overflowed and flooded the city Borovsk, including the lower floor of the Tsiolkovsky home. Everything in that part of the house was ruined. The older Tsiolkovsky daughter, Lubov, related the following in her memoirs:

> Father threw a bunch of logs on the floor and placed wood boards over them and used it as a bridge for us to get our belongings and then we floated along the top of the water. We went upstairs. What was frightening was watching through the window ice floes float by outside. It was cold and scary and we could not sleep.

The water quickly subsided. But the flood caused a dampness that the wind and spring thaw could not dry, so we decided to again move.

Now the family moved to the center of the city. A friend and fellow teacher, Evgeny Sergeevich Eremeev, helped them with establishing themselves at a new residence. (The street is now named Tsiolkovsky St.) Daughter Lubov recollected the move:

> We were children and we became happy with the new apartment with three rooms. It was such a large home and the largest we lived in. What was most important was that it was made of stone, so it would not burn down, and far from the river with less chance of flooding.

Another room also came with the apartment, and this became Tsiolkovsky's workshop and library, but the disadvantage was its cost. At six rubles a month, this was a quarter of his income as a schoolteacher.

Now that they were living in an accessible area instead of at the edge of the city near the river, visitors started to circulate through the Tsiolkovsky home. Lubov recollects the period:

> My mother would entertain visitors. I remember my complete bewilderment when a Borovsk resident, overweight, in a high-collar shirt and starched cuffs, polished boots, would sit in one of our chairs, ask about the health of each of the members of our family, stay for about 5 minutes and then leave.

Guests such as these were busybodies with nothing else to do but interfere with Tsiolkovsky's work and squander his time.

Tsiolkovsky continued on his dirigible design and in 1890 published his next article, *Of the Possibility of the Construction of a Metallic Aerostat.* In 1892, he decided to send it to the person who would do the most on behalf of Tsiolkovsky, Dmitry Mendeleev. With the treatise, Tsiolkovsky enclosed in the package a small model of his fixed-shape design, but made of wood and paper and glued together. Mendeleev forwarded the treatise and model to Vyacheslav Izmailovich Sreznevsky for further consideration. In his letter, Tsiolkovsky requested the society to provide him assistance both financially as well as scholarly. He asked for 300 rubles, a healthy sum considering that Tsiolkovsky only made 24 rubles a month as a schoolteacher, but affordable for the society considering their financial status. However, some of the members again categorically rejected the concept of a metal-clad dirigible filled with a lighter-than-air gas. The primary opponents to Tsiolkovsky's design were the high ranking military engineer Evgeny Stepanovich Feodorov and general-lieutenant Aleksandr Matveevich Kovanko, who was also director of the aeronautics school in Petersburg.

At his school Kovanko had done some research on his own in earlier years and he had come to the conclusion that a metal-clad aerostat was not a viable concept and would not succeed. His inclination was toward heavier-than-air vehicles. (The Russian aviation pioneer Aleksandr Mozhaiski built an airplane that made its initial flight in 1884, although hardly considered successful. But Mozhaiski was part of the technical elite of Imperial Russia, being an admiral in the navy, so his design was pursued instead.)

At a meeting of the Technical Society on October 23, 1890, Evgeny Feodorov complimented Tsiolkovsky's zeal and his dedication to aeronautical research and experimentation, but said that the concept of a metal-clad dirigible was unfeasible in practical application. This left Tsiolkovsky with a sugar-coated bitter pill to swallow. They also stated they had evidence of

European and American designs of metal-clad dirigibles with no successful results after placed into service. Of course, Tsiolkovsky was also denied any financial subsidy.

It did not help Tsiolkovsky that subsequent newspaper articles on the conclusions of the society's meeting misspelled his name as Tsankovski. The articles stated the dangers inherent in a metal-clad gas-filled dirigible, that the thin metal leaves connected together to create the shell could fail at any time from either wind turbulence or internal pressure, or just any kind of vibration.

A few individuals, out of pity for Tsiolkovsky's plight, did forward him some funds, but hardly enough to invest in anything large scale. To defend himself, Tsiolkovsky wrote letters to Stoletov and Feodorov and a couple of other members of the society, but these were a dead end.

Tsiolkovsky continued his research, and in 1891 completed his first complete book on the subject, titled *A Navigable Metal-Clad Aerostat*. He sent it to a Moscow publisher, M.G. Volchaninov, for printing and distribution.

Then all of a sudden Tsiolkovsky packed up the family and moved to Kaluga. The reason, based on indirect information, was a conflict between himself and other teachers and some of his students' parents. The conflict arose because Tsiolkovsky discovered that some of the parents were bribing teachers to pass their children through school and grant them diplomas when they had not met grade requirements. Tsiolkovsky felt this was unethical and voiced his concern, but due to repercussions from city officials for exposing the misconduct, and their desire to remove Tsiolkovsky from his teaching position in revenge, he concluded it best to move to another city. He would never return to Borovsk but remained in Kaluga to the end of his life. At this time it was a city with a population of about 50,000.

Matters did not turn any better for Tsiolkovsky with the Technical Society, which categorically rejected his letters again and his book. They stated that any dirigible based on Tsiolkovsky's design would have a sub-standard construction and be liable to failure.

Nonetheless Tsiolkovsky continued on his research and he published a second volume on his designs in 1893, an addendum to *A Navigable Metal-Clad Aerostat*, where he also answered and resolved some of the comments made by critics. His primary concentration in this second volume was the design of a steering system, as opposed to hot-air balloons that were at the mercy of the wind. His designs included a passenger department to hold up to 200 people and some cargo, a means of balancing the dirigible to keep it level, a rudder with a wheel to control it, a gasoline motor in the control room driving a generator to operate the steerage and control the velocity. Mercury float switches would be used to indicate any imbalance in the dirigible. A cylinder over the length of the passenger compartment would be filled with water. As the dirigible tipped forward or backward the mercury switch would trip pumps to move the water into another area in the cylinder to stabilize the dirigible. Tsiolkovsky also felt that the dirigible's balance, speed and direction could be made to operate automatically. Tsiolkovsky was 20 years ahead of his time in the design of an automatic steering system, far ahead of Elmer Sperry.

On January 15, 1893, E.S. Feodorov again presented a report to the Technical Society on Tsiolkovsky's dirigible design, and again stood firm in his conviction of the impracticalness and unfeasibility of such a design, even with the corrections and supplements in Tsiolkovsky's second volume on aerostat design and steerage. Feodorov though believed in Tsiolkovsky's sincerity and used all possible means to inform him of his miscalculations, hoping to steer his effort toward what the society already considered practical and feasible. In addition to Kovanko, a third opponent joined with Feodorov, Artillery General Mikhail Mikhailovich Pomortsev, another academic who was convinced that a dirigible was unsteerable and subject to atmospheric forces. Both Zhukovsky and Mendeleev also changed their attitude toward Tsiolkovsky at this time, aligning themselves with the dominant opinion of the society members, mainly due to pressure from Mozhaiski.

By this time Zhukovsky had not yet developed his theory of the force of jet propulsion and was still investigating the topic, but the flow pattern proposed by Tsiolkovsky seemed for Zhukovsky to be closer to the truth than what Newton had proposed. Zhukovsky recorded his opinion of Tsiolkovsky's most recent book in a letter to Stoletov:

> The treatises of Mr. Tsiolkovsky cause a pleasant impression, since the author, utilizing a small amount of analysis and inexpensive experiments, came for the most part to reliable results. Although the majority of such results are already well known to the scientific world, but nonetheless the original method of the author's research, deliberation and precision of experimentation do not curb interest in his conclusions and in every case characterize him as a talented researcher.

The rupture between the worlds of Feodorov and Tsiolkovsky can be better sensed with a familiarity of the matter regarding David Schwartz. This matter started about 1892, when Corporal Zuev, a Russian military agent in Austria, informed his superior officer, a minister of defense, that the Austrian inventor David Schwartz proposed to build equipment to steer air balloons. The minister of defense invited Schwartz to come to Petersburg and build an experimental unit.

What made this matter embarrassing was that the committee having control over the work of the Austrian inventor included Tsiolkovsky's adversaries: M.M. Pomortsev and A.M. Kovanko. But Schwartz had no credible design or calculations or experimentation to prove his aerostat would work, and the attempt was a massive failure. The shell of his manufactured aerostat deformed while being filled with gas and the aerostat burst open, releasing all the gas.

Fleeing Petersburg to Germany, Schwartz received some support from the owner of an aluminum factory. Having gained some experience from his failure in Russia, he built a new model aerostat. Schwartz died shortly after, but when the final product was released into the atmosphere, what occurred was hardly called a success, from the testimony of witnesses. The design was shelved for a while until Count Ferdinand von Zeppelin, a German retired military officer, purchased the rights to the design and then later in 1900 constructed and flew the first steerable German airship, now called a zeppelin. It had an aluminum frame covered with fabric, with gas-filled balloons inside.

Naturally, this history did not leave Tsiolkovsky indifferent. He accused Schwartz of being an imposter, but there was little that Tsiolkovsky could do, a small-time educator in a small Russian city.

It is difficult to suppose that Schwartz knew nothing about Tsiolkovsky's aerostat designs and published treatises. Correspondence is available proving Schwartz's communication with Kovanko and other individuals who were well familiar with Tsiolkovsky's efforts at the time. The printing presses went rampant about the successes of Schwartz in Germany and the subsequent successes of Zeppelin with a full-size working model. But they ignored the fact that the Germans used Tsiolkovsky's design on a major scale, and there was nothing he could do.

Russian engineers, however, shelved the design of aerostats and proceeded on heavier-than-air flying machines, which design was supported as more feasible and practical by Tsiolkovsky's rivals.

As far as the publication of his latest treatise was concerned, *A Navigable Metal-Clad Aerostat*, only 13 copies were sold over the years 1892–1893. This further disappointed Tsiolkovsky.

Reading in technical journals about the bold attempts of flying, or better stated as gliding, by Otto Lilienthal, and also about Hiram Maxim, who also attempted to construct an airplane, Tsiolkovsky did not feel he should be left behind in this area of aeronautics. Step by step Tsiolkovsky proceeded to analyze the possibility of a heavier-than-air flying machine and he utilized his earlier studies on aerodynamics and aerostat design in this new direction. In 1904, he published his treatise, *The Aeroplane or the Bird-like Flying Machine.*

He claimed that earlier, in 1900, he already had preliminary calculations and designs completed for such a vehicle. Tsiolkovsky was familiar with the many previous attempts with flapping wings, but did not believe any such design would be practical, since none of their attempts at long distance flight were successful. It was not so much that flapping wings required a complex construction, but to cause them to flap enough to acquire up-lift and flight required a tremendous amount of energy, and this was unfeasible.

Hiram Maxim's airplane utilized a steam engine for the propellers, but the immense weight as a result made it a poor design. By this time, Maxim abandoned his efforts.

Speed, as Tsiolkovsky calculated, had its positive effect in regard to force of up-lift, but a negative effect in regard to air resistance. The faster the airplane, the greater the lift and the greater the resistance. As with his other innovations, Tsiolkovsky proceeded with blind faith in this design. Two propellers at the front of the aircraft would provide the necessary speed for draft for flight, while for vertical and horizontal movement Tsiolkovsky design tail wings similar to those on a bird. The source of energy would be a gasoline-driven engine that would be cooled by the stream of atmospheric wind. The tail wings would be operated from the seat of the driver. Tsiolkovsky utilized the seagull as his prototype, feeling this bird had the best relative dimensions and size for a monoplane, as a result of a combination of speed and minimal air resistance, and wing design. But Tsiolkovsky's envisioned monoplane did not materialize into production for over 20 years.

As with the aerostat, his adversaries at the Physics-Chemistry Society discredited Tsiolkovsky's idea of an aeroplane. They could not fathom the former, claiming a metal-clad shell would be too heavy to fly, and the latter concept based on the seagull using a gasoline engine and fixed wings for speed and uplift likewise astonished them.

Other heavier-than-air constructions utilized bamboo as a material to keep the vehicle as light as possible. Tsiolkovsky's design was not based on weight but rigidity, so his construction used a truss design covered with fabric for the wings and fuselage. (The Eiffel Tower gave him the idea.) For some reason, Tsiolkovsky's effort in the direction of an aeroplane design was diverted at this time and he returned to his earlier research on aerostats.

Tsiolkovsky's research conclusions on the aerostat by this time were stunning, but he also realized its complexities in real application. He wrote in one of his treatises:

> Based on all this it is evident that the construction of such a vehicle to contain a significant amount of air travelers is difficult. Nevertheless, my larger theoretical aerostats, under the best of conditions, are still

> able to contain as many as 600 passengers and could have a speed as high as 40 miles an hour. This is still 100 times less energy per person than conventional means of travel, and a 100 times greater possibility to construct and a 100 times less expense per passenger for travel.

Again to confirm the feasibility of his convictions, Tsiolkovsky continued to publish his treatises, the next one in 1894. But as before, the members of the Petersburg society did not turn any further attention to his latest research.

In 1896, M.M. Pomortsev further explained his opposing views of a metal-clad gas-filled dirigible in his book, *Restraints of a Free and Steerable Aerostat.* Pomortsev claimed a study of 40 flights of cloth-covered dirigibles with gas-filled balloons and included his calculations and research in this book. After its publication, Tsiolkovsky read it and discovered several incorrect formulas and calculations and conclusions and which caused Pomortsev's logic to fail. Tsiolkovsky then published his rebuttal in the journal, *Technical Collection.* Upon reading Tsiolkovsky's article, Pomortsev literally tore the hair from his head, and he ran to all the booksellers gathering all the unsold copies of his book, but was unable to gather them all. This forced the Petersburg Society to issue an apology and some gratitude to Tsiolkovsky. Later that year, in the November issue of *Kaluzhki News*, E.S. Feodorov composed an article on the future of dirigibles and concluded his article with the statement:

> The work of Mr. Tsiolkovsky evidently is the fruit of a solid effort expressed very equationally and has its merits, even among his opponents.

Tsiolkovsky took advantage of the situation to request additional experiments of model aerostats to determine actual wind resistance, its ability to fly, and its navigational capabilities. These experiments were done outside in the open and depended on regular atmospheric wind. Tsiolkovsky had a method of determining wind speed for all his calculations. The terminal goal of his research left no doubt remaining: he measured wind resistance against his model dirigibles and determined formulas to use in applications of large scale vehicles. Tsiolkovsky published his results in a brochure with the lengthy title, *An Iron-Clad Steerable Aerostat for 200 Persons, as Long as a Large Steamship*. Some of it was a repeat of his earlier treatises of 1891–1892, with his design for navigating and balancing the vehicle, however, now improved. At the same time, the results of these experiments were also applied to airplanes.

What was quite edifying with the results of the experiments was that they matched and coincided with the calculations of earlier years made by Tsiolkovsky and this caused the technological community to take further notice of the feasibility of a metal-clad dirigible with Tsiolkovsky's design for propulsion, steerage and balance. In another respect, the results did not bring Tsiolkovsky much closer to the members of the Society, since his calculations proved wind resistance to be much lower than calculated by Pomortsev and others, and this was the determining factor of the feasibility of such a vehicle. Tsiolkovsky also proved that the faster the vehicle, the less the relative resistance. Now the dirigible could sail considerably faster with less energy consumed, making it a viable project for full-size development. Tsiolkovsky now considered himself the victor in the technological struggle.

The experiments needed to continue in order to refine the initial results, but he could no longer depend on natural wind for resistance testing. Now he needed a wind or aerodynamic tunnel, but this would cost some money, hundreds of rubles. Of course, Tsiolkovsky decided he would construct his own. With his personal finances in a state of near destitution, a wife and seven children to support, Tsiolkovsky turned to the Chemistry–Physics Society of Petersburg asking for 200 rubles.

The society responded by asking for further details on the use of the funds to determine whether they should subsidize him, they claimed their account was small. Tsiolkovsky's response was the possible use of dirigibles in military maneuvers, yet he wanted to keep as much as possible a secret, lest some of his plans migrate to some country who would use it in a military intervention. The society responded by requesting more information on his design of a wind tunnel. Of course, Tsiolkovsky replied with a general description: a motor with fan inside of a long tube, with measuring instruments located in proper places.

As in the past, the society informed Tsiolkovsky that his further experiments with a wind tunnel of the construction and size he described would not succeed. As far as the society was concerned, in order to provide results that would be sufficient to interest them and be of practical use to science, the wind tunnel would have to be designed on a much larger scale, and the society did not have the funds to build one of this size.

Tsiolkovsky was embittered; he needed financial support now more than ever. However, he managed to build a wind tunnel and at the expense of his family's subsistence: depriving them of domestic needs. For all practical purposes, Tsiolkovsky constructed the first wind tunnel in Russia for use in aerodynamics. The fan was driven manually. As a result, Tsiolkovsky pioneered the era of experimental aerodynamics.

Tsiolkovsky experimented with about 100 additional sizes, lengths and shapes of dirigible models, all of them fabricated from heavy drawing paper glued together over a wood frame. (Not one of them survived, being destroyed in a flood in the region in 1908, although Tsiolkovsky did take some photographs of them.) These experiments did not provide the results that Tsiolkovsky wanted because the equipment was undersized; he needed to increase the size and shape of his wind tunnel to provide the volume required for the experiments, and he needed to install a larger fan to increase wind velocity. The same problem again plagued him: lack of money. As a result, Tsiolkovsky terminated his experiments. Nevertheless, he utilized to the best of his ability the data that he did acquire and further developed mathematical equations related to questions dealing with aerodynamics.

On May 10, 1897, Tsiolkovsky derived the formula that established the relationship between the velocity of a rocket at a specific moment, the mass of the rocket, the mass of the burning fuel, and the velocity of the gas from the exhaust nozzle. This became known as the Formula of Aviation. Of course at the time, Tsiolkovsky had no idea of the importance this equation would have for future designs of jet propulsion equipment.

In 1898, Tsiolkovsky published his treatise, *The Pressure of Air Introduced in an Artificially-Induced Air Stream on a Surface*, in an Odessa journal.

Tsiolkovsky's experiments appeared to be crude and incomplete, but the appearance is deceptive. In reality, they were exceptionally precise and accurate. He was the first in the history of experimental scientific research to determine the impact of a force or resistance of a stream of air at various velocities on the outside surface of a moving body. The observant scientist noticed the role played by the resistance of the stern of the moving body, as opposed to the conclusions of M.M. Pomortsev who vehemently dismissed that. Tsiolkovsky was absolutely certain of the accuracy and correctness of his conclusions. As an energetic champion of experimentation, he was ready to repeat any experiment, guaranteeing reaching the same precise results. He reflected on his experiments in later years:

> The instruments that I built were so inexpensive, simple and easy to use, and they quickly resolved irresolvable theoretical questions that some counted to be the sole custody of some university or their physics' department.

Tsiolkovsky shortly after sent a letter to the vice-president of the Petersburg Academy of Sciences explaining the concepts and conclusions that were earlier published. Expecting as usual a letter of rejection, Tsiolkovsky was surprised when 10 days later, on September 22, 1899, the Chemistry–Physics Department invited Tsiolkovsky to meet with one of its most distinguished members, Mikhail Aleksandrovich Rikachyov, and who was several years older than Tsiolkovsky. It seems Rikachyov had likewise been interested in the concepts of air flight for many years and attempted himself to construct, although in secret, some flying vehicles in various designs. He was convinced that air travel would surpass travel by land or sea, once the question of propulsion was resolved. In earlier years, 1870–1871, Rikachyov had also conducted some personal experiments in aerodynamics, but kept his results personal. With Tsiolkovsky, Rikachyov seemed to have found a serious and cooperative researcher.

Rikachyov was also to investigate Tsiolkovsky's plans for future experiments and determine if they were worth funding by the Society. Within a week after meeting with Rikachyov, Tsiolkovsky wrote to the society:

> I cannot list in detail every item of my agenda and the time and cost of each, but I think that 1,000 rubles will be sufficient. Nonetheless I will be grateful even for the least amount the Academy of Sciences will be willing to contribute. Do not disdain my humble request because the thought that I am not alone in this provides me moral strength to speedily advance my intended efforts and with the help of God to be able to conclude them by autumn 1900. Photographs, sketches and all scientific data will be sent to you at the conclusion of my experiments.

Even in view of the involvement of Rikachyov, the Academy only forwarded 470 rubles, and even then, such a sum, as meager as it was compared to what Tsiolkovsky actually needed, overjoyed him. Now with the support of the Academy of Sciences Tsiolkovsky was convinced he achieved indisputable victory regarding the feasibility of his gas-filled metal-clad, balanceable and navigable dirigible. Tsiolkovsky went to work immediately on a large-scale aerodynamic test tunnel, and as always, once that he would design and construct himself. At the same time, Tsiolkovsky was developing his automatic pilotage based on gyroscopic control.

It took Tsiolkovsky almost a year to build his new wind tunnel, and the second year was dedicated to his experiments. What Tsiolkovsky did not have was an electric motor-driven fan to use to propel air through his tunnel, so the fan was driven manually.

Tsiolkovsky still felt deep inside of him that the funds were routed to him by the Society not with the intention of him proving his dirigible feasible on a large scale, and so promoting its construction, but hoping that his conclusions would prove otherwise. Somehow the Academy was not going to allow a deaf and almost destitute, high school mathematics teacher without a formal education and from some outlying province, to get the better of them. This failure could be further used by the Society to discredit his concepts of inter-planetary and outer space travel by jet-propelled rocket ship. Even by this time, not one full scale dirigible of any design was constructed in Russia, although many were in Europe and America.

The stress on Tsiolkovsky between family, work and research had a disastrous effect on his health. He almost died of peritonitis. This did not dampen Tsiolkovsky's efforts and a local friend assisted him, Vasily Ivanovich Assonov, who was otherwise employed in a position similar to a county inspector. Another was Professor Peter Lavrovich Lavrov, an inquisitive person likewise drawn to Tsiolkovsky because of his interesting endeavor. Lavrov saw Tsiolkovsky's book *An Iron-Clad Steerable Aerostat* in a library and read it. Motivated, he went to Kaluga seeking the author and then remained there for the next 26 years until his death.

A.B. Assonov, a son of Vasily, relates how his father invited Tsiolkovsky to their home and shares his impression:

> On the following day, someone rang the doorbell (it was manual at the time). I opened the door and told my father that someone had arrived. I remember it as though it happened today: a person entered in an autumn coat, taller than average, long hair, eyes that were black and blacker. Under his coat was another long coat with short sleeves. During their conversation he became agitated and turned red. Father invited him to the living room and they sat near our piano. They spoke for a long while about his publications and work. I stood at the door and listened. Quickly Tsiolkovsky left, upset, while putting his coat on. Father then told me that this teacher is the famous mathematician and they had to exert all effort to publish his new work to the scientific community. Shortly after, the second edition of his book *Aerostat* was issued.

Tsiolkovsky continued composing his scientific treatises: *Is a Metallic Aerostat Possible?*, and *Gravity as the Source of Universal Energy*, in 1893, and *The Longevity of the Sun's Rays*, in 1897, all published with the help of Assonov.

About this time, V.I. Assonov put Tsiolkovsky in contact with the president of the Nizhni-Novgorod club of amateur physicists and astronomers, Sergei Vasilyevich Scherbakov. They corresponded regularly for about the next 20 years. This club provided Tsiolkovsky with some financial support over the years.

In April 1893, Tsiolkovsky applied for membership in their club, feeling a friendship with the president and other members with whom he corresponded. He was accepted unanimously in December of that year. But as with his membership in the Petersburg Society, Tsiolkovsky had no spare funds with which to pay his dues, and as a result he asked to be excused from membership. However, the club refused his resignation and continued to keep him as an honorary member.

Tsiolkovsky's treatise *Gravity* was first published in one of the club's circulars, and then later in the journal *Science and Life*. The editor of this journal was a doctor, N.M. Glubovsky, and he willingly published Tsiolkovsky's submissions. There were 500 copies made of the edition of *Science and Life* that included Tsiolkovsky's treatise *Gravity*, and he hoped that some of the income could be used for membership dues. But hardly enough were sold to begin with. The demand for such scientific compositions was limited. Other journals would not print these treatises and Tsiolkovsky had to settle for only a small advertisement pointing readers to some bookseller where they could be purchased. Essentially, Tsiolkovsky received hardly a kopek from their sales after costs of publication and sales were accounted for. Later the balance of remaining magazines and brochures were distributed for free to interested parties, whether Tsiolkovsky knew them or not. Regardless of such circumstances, Tsiolkovsky continued his research and compositions and experiments.

Following this event there occurred a social disaster for Tsiolkovsky, a scandal that caused him much irritation. It involved A.N. Goncharov, owner of a large estate near Kaluga, very educated, a university graduate, conversant in various foreign languages, employed at a local bank, and nephew of the famous Russian author Ivan Alesksandrovich Goncharov.

The relationship began in a favorable direction when Goncharov read some of Tsiolkovsky's treatises. He became so enthused over the possibility of such flying machines that he had his wife, Elizabeth Aleksandrovich, translate a couple of Tsiolkovsky's brochures into French and he sent them abroad. The results of their opinions on Tsiolkovsky's concepts were very positive, with some articles published in French newspapers in 1897, and this was encouraging.

Sometime in 1894, Tsiolkovsky read selections of his compilation of science fiction stories with some of his cosmic philosophy, *Dreams of Earth and Heaven*, to a group of people gathered at the home of A.N. Goncharov, and this included his short story *On the Moon*. The contents interested Goncharov and he offered to help have the book published. The book, however, had to be approved by the government censor and this delayed publication. Eventually the book was approved and it went to print.

Goncharov did not know that Tsiolkovsky had planned a surprise for him. For assisting Tsiolkovsky with the publication of the book, when the first copies came off the printing press Tsiolkovsky printed on the cover in large letters the statement, "Published by A.N. Goncharov," as a sign of gratitude. Tsiolkovsky took a copy to Goncharov who, upon seeing it, did not appreciate it at all. He went into a rage and threw Tsiolkovsky out of his house. Goncharov felt insulted that his name would be associated with a bunch of science fiction nonsense by an outer-space dreamer from a second-rate provincial city, and now Goncharov would be labeled as vain and fame-seeking.

Tsiolkovsky had hardly begun to forget about the offence taken by Goncharov when a new humiliation was thrown at him (and in this case by a reviewer who refused to state his name) in regard to Tsiolkovsky's collection in *Dreams of Earth and Heaven*. This was published in the journal *Scientific Review* in May 1895:

> We would willingly call Mr. Tsiolkovsky a talented popular writer and if possible a Russian Camille Flammarion if this author possessed some sensibility and did not have intents of removing the laurels from Jules Verne for his personal advancement. His selected book makes a sufficiently strange impression. It is difficult to guess where the author is serious or where he is fantasizing or where he is just joking.[1]

> If the scientific explanations of K. Tsiolkovsky are not always sufficiently based, the projectile of his fantasy is positively intolerable and even transcends the nonsense of other science fiction writers who in any case still have more of a scientific basis than does Tsiolkovsky.

> So we have an author here with some kind of space creatures or residents of asteroids who agree to create a circle[2] or triangle to direct rockets as carriages, moving them arbitrarily close to the sun.[3]

It was apparent that the reviewer's comments were intended to be malicious or at least to humiliate Tsiolkovsky, since his intent was to motivate people through science fiction to think about space travel and the means of accomplishing it, a mental exercise which seemed to be beyond the reviewer's capacity. Tsiolkovsky also used his research and the conclusions of his experiments to justify descriptions of the experience his space travelers and his cosmic philosophy.

About the same time, other Russian science fiction writers likewise published books about space travel and the alien creatures that could inhabit other planets, such as G. Altov and V. Zhuravlyova. Their imaginations went far beyond Tsiolkovsky's, since their works were purely fantasy while Tsiolkovsky combined his with scientific facts.

Nevertheless, the bitterness of this rupture with Goncharov and the humiliating comments seem not to have reduced Tsiolkovsky's interest in cosmic matters. In 1896, the newspaper *Kaluzhski Messenger* published another of Tsiolkovsky's articles. This latest article dealt with the possibility of actual communication with outer space and creating cosmic ties with other planets and solar systems.

It was not without difficulty that Tsiolkovsky finally concluded his experiments with his home-made manually-operated wind tunnel. However,

1 Referring to his cosmic philosophy, some of which is included in this volume.
2 The Russian word used here also means club, so the author is making a humorous play on words with his subsequent use of triangle.
3 A reference to Icarus who flew too close to the sun and the wax on his wings melted.

given the amount of opposition blocking him from the construction of a full-scale model, he set the results of his tests aside and returned to his first love: the exploration of the cosmos.

In November 1896, Tsiolkovsky started his short science fiction novel *Beyond Earth*, a narrative of the rocket space flight of several scientists from their respective countries to several planets, but he only got as far as Chapter 10. He continued working on it in January 1917, finishing all 58 chapters in 1918.

In 1903, Tsiolkovsky expounded his thoughts in a large, serious work titled *Exploration of Outer Space Using Jet-Propelled Rockets*. This latest composition enjoyed only a very limited readership. People seemed to be more interested in fiction than the actual possibility of space travel based on calculated scientific evidence.

Suppression of Tsiolkovsky's efforts by the Petersburg elite did not go without notice. On October 11, 1897, the *Kaluzhski Messenger* published a short article titled *No Prophet in Our Fatherland*. It dealt with the unnecessary and forced silence by Russian scholars on Tsiolkovsky's efforts. The article agitated at least one enthusiast of Tsiolkovsky's work, Pavel Mikhailovich Golubitzky. He visited Tsiolkovsky at his home in Kaluga.

Tsiolkovsky explained to Golubitzky his struggles, plans and intentions, and the obstacles that would arise to interfere with his research that was so necessary for the generation of a genuine scientific approach to rocketry. Golubitzky paid attention and took the time to become familiar with his home-made aerodynamic laboratory, looking at the crude and poor quality instruments. Stacks of data and unpublished binders of research papers were scattered all over the room. It was clear that Tsiolkovsky had succeeded in accomplishing a great deal and could have done more and better had he the proper instruments and facilities, and he still could in the future.

Hardly a week passed before a lengthy article appeared in the *Kaluzhski Messenger* bearing the large heading, *About Our Prophet*. It was a resolution to the problem stated in the earlier article. Part of the article is the following:

> When I left Tsiolkovsky's home I was burdened with heavy thoughts. On the one hand I thought, "This is the 19th century, the century of great inventions and discoveries, a passing stage, just as Stoletov prophesied, from the age of electricity to the age of ether. And on the other hand, the absence of any possibility for a destitute laborer to become familiar with all the efforts of those persons who would interest him."
>
> Such deprivations over the years will only cause Tsiolkovsky to develop and die from tuberculosis, and after his death there will pass perhaps a hundred years, or who knows how long, until another

> inventor will be born, one with a capacity of self-abnegation for the sake of his purpose, who with his own efforts will cause that moment to approach, when people will rush through the atmospheric ocean just as at present they drive over the land's surface.
>
> I turn to you highly respected professors and titans of Russian science. Show your mighty support of a destitute laborer whom you would label a black slave. Throw him some crumbs, help him with your counsel. I turn to you people who seem to be estranged from his scientific effort and announce to you that many competent people have recognized the great scientific significance of Tsiolkovsky's effort, and so you have the obligation to assist him.
>
> I ask you, editors of Russian newspapers and journals, in the interests of the capabilities of Russian inventors, to not reject him when he asks you to publish his research notes and treatises.

This passionate defense of Tsiolkovsky's activities served its purpose and a ray of hope shined into his house. Matters now started to turn for the better. The Kaluga intelligentsia ceased treating Tsiolkovsky as an alien creature. The forefront of its citizens realized that they were actually dealing with a serious scholar and scientific thinker, although self-taught, and now viewed all his books and publications as a genuine contribution to and advancement of Russian technology.

Next was a complete turnaround in the attitude of the Kaluga population toward Tsiolkovsky. For several years Tsiolkovsky had been teaching mathematics and geometry at a private school and was doing an excellent job. In December 1896, he received an award. Then in 1899, Tsiolkovsky received a promotion with a transfer to teach mathematics at a provincial girls' high school. It was unprecedented for a person without a diploma from a higher institution to teach at this type of school, but Tsiolkovsky became the exception. The district director of schools, P.A. Rozhdestvensky, stated the following regarding Tsiolkovsky's promotion:

> Mr. Tsiolkovsky, an instructor of arithmetic and geometry, is a comprehensive specialist in his subjects and teaches them with a special capability: clarity, precision, determination, a strict discipline and presentation, with a natural ability in explaining and providing lessons in mathematics.

Tsiolkovsky greeted the 20th century with the encouragement of several successes. Regardless of all the obstacles that fell directly in his path, he always believed in the future since the universe was infinite and eternal. Nonetheless, the next few years brought him some sorrows.

Tsiolkovsky's experiments brought interesting results. He succeeded in determining formulas still in use to the present day, regarding the aerodynamic coefficient of the resistance of air relative to the velocity of a body. Another set of calculations dealt with laminar flow at low velocities and turbulent flow at high velocities and the resistance of the liquid on the surface of the body. Later this was used in hydraulics. They were completed in 1902.

The usefulness of his completed work was beyond doubt, but the Academy did not seem to think so. Rikachyov explained his reasons for not wanting to include Tsiolkovsky's works in the Academy's publications as follows, and this was just an excuse so not to accept the validity of his experiments:

To resolve the question of the inclusion of the works of Mr. Tsiolkovsky in the publications of the Academy of Sciences, it is absolutely necessary as a preliminary consideration to request from the author the records of all his observations in an organized form, grouped in a manner so that each conclusion noted in the text was proven by the observations upon which this conclusion was based, with an indication of at least the date when these experiments were performed. None of the observations that do not agree with the results should be excluded and a reason should also be provided as to why such observations did not agree with Tsiolkovsky's final conclusion. The raw data must also be included with the method of experimentation in a table.

How can the unscrupulous character of Rikachyov be described, demanding a complete description of Tsiolkovsky's experiments, data, observations, and results all in tabulated form? What first came to Tsiolkovsky's mind was that Rikachyov was accusing him of falsifying raw data and the conclusions derived from them. But this was not all; Rikachyov also intended to blatantly exclude Tsiolkovsky from the papers of the Academy of Sciences. Tsiolkovsky was insulted and refused to comply; and the Academy refused to publish his extensive volume. Only a few excerpts were later published in the *Scientific Review*. Tsiolkovsky was glad that at least the Academy did fund some of these experiments, and all his conclusions were proven true.

An irrecoverable tragedy then struck the Tsiolkovsky family. In December 1902, Tsiolkovsky's eldest son Ignaty died in Moscow. It was a suicide. The incident psychologically devastated Tsiolkovsky and for several years he was unable to forget or overcome the horrible loss. He recorded the impact of the tragedy later in his memoirs:

> A new stroke of fate impacted me in 1902: the tragic death of my son. Again a horribly remorseful and depressive period for me began.

As soon as I would wake in the early morning, I already felt empty and troubled. It was not for 10 years after the event that this feeling subsided.

I placed the welfare of my family and associates on the lowest plain, with my aspirations on the highest. I did not drink, did not smoke, did not squander even one extra kopeck on myself, for example, on clothes. I was almost always hungry, poorly dressed. I was moderate in all items to the extent possible. My family endured all of this with me. I was often vexed and perhaps as a result this made life difficult and nerve-wracking for those surrounding me.

Upon their return from Moscow, Tsiolkovsky and wife Varvara would sit long hours in private, comforting each other. Daughter Lubov related this in her later memoirs. The couple then decided it would be best for the family to seek a better home instead of the cramped flat they had in Kaluga, now that they had more income available from his new teaching position.

In late 1903, Tsiolkovsky's latest treatise was published with the lengthy title, *The Internal Pressure of the Sun and its Compressive Effects based on the Elasticity of their Material*. Now Tsiolkovsky's relationship with publishers was improving. His next treatise, *Successes of Air Flight in the 19th Century*, was a review of books and attempts of air flights compiled by other authors on the subject, and *The Resistance of Air on Air Flight*. Both of them were published in *Scientific Review* and Tsiolkovsky was now recognized as a regular and valued contributor as well as becoming a friend of the publisher M.M. Filippov.

Essentially the cosmic era, meaning the era of space travel, began with the publication of Tsiolkovsky's *Exploration of Space Travel* in 1903. Tsiolkovsky unfolds his thoughts logically and convincingly. Tsiolkovsky's earlier works as well as those of other authors on this subject were not definitive or as scientifically based as this one, or were primarily science fiction. Likewise the new method of propulsion was something not theorized earlier: it was jet propulsion. It was a motive force different than that applied to aerostats, airplanes or hot air balloons, or even others such as H.G. Wells' cannon which he envisaged to shoot a projectile into space for his first "men on the moon."

Tsiolkovsky's first argument in his treatise presented the aerostat as a poor candidate for space travel, since it would not be able to rise high enough. As the aerostat increased in altitude, the density of the external air decreased and so the internal gas volume would increase in pressure, which would in turn expand the shell, eventually exploding at a sufficiently high elevation.

The second poor candidate was a projectile from a cannon. For most of history, the sole means of ejecting a projectile had been with gunpowder. That method would be difficult to utilize for space flight, since the vehicle would have to leave the ground at such a high velocity, relative to its weight and size, that it would destroy itself at the moment of discharge (if not shortly after some ascent). Tsiolkovsky calculated that a gun barrel of about 1,000 feet in length would be required, under the most massive amount of explosives, while the highest altitude the projectile would reach would be about 200 miles. Then it would stop due to the gravitational pull of Earth. The subsequent pressure on a passenger would be murderous. This was unfeasible.

The resolution was jet propulsion. This concept set Tsiolkovsky apart and made him distinct from all his predecessors and contemporaries.

Tsiolkovsky's proposal was to use hydrogen and oxygen as a fuel that could be carried onboard the vehicle. The calculations he presented indicated promising results. Their blending and ignition at a steady rate would produce propulsion sufficient for the purpose of resisting Earth's gravitation force, and as the rocket rose and the gravitation force decreased, the motive force would also proportionately decrease. The steady velocity of ascent would not impact the passengers' safety or health. The size of such a vehicle would not restrict its movement, designed in an oval shape. Tsiolkovsky calculated the force necessary to lift the rocket beyond the effect of Earth's gravity with a basic design of the vehicle. What was lacking to complete Tsiolkovsky's rocket was the actual amount of fuel necessary to create enough motive force or propulsion to accomplish all of this.

In addition, Tsiolkovsky included an automatic control for steerage of the rocket and one which he felt was very practical, which he borrowed from those he had applied to his earlier designs for the aerostat and airplane.

Tsiolkovsky likewise described the possibility of concentrating the sun's rays and utilizing the sun as a pilot for the spaceship. If the course of the spaceship could be fixed relative to the course of the sun, then any deviation of the spaceship from its course would be noticed by the controls which would then bring the spaceship back on course. The control would release some of the gaseous mix and an electrical spark would ignite the fuel, causing the spaceship to be propelled back onto its proper course; and then the flame would stop, the spaceship floating as though in neutral gear.

Tsiolkovsky did his best to calculate the spaceship's ability to move goods from place to place, meaning planet to planet and star to star, and that this was the most economical method. Although its ascent from Earth would be slow at the start, the spaceship would gradually increase in speed due to the decrease in gravitational force and this would propel it into the distance reaches of outer space at a high velocity, allowing short periods of travel time between destinations. Since the environment is a vacuum with no air resistance or gravitational pull, travel would occur without propulsion, except to change direction. Tsiolkovsky likewise analyzed the behavior of the rocket outside the atmosphere in a region unaffected by any gravitational force.

Tsiolkovsky was so distant, ahead of his era and his contemporaries, in his vision of future rocketry and space flight! Tsiolkovsky mentioned in his memoirs:

> This work of mine is far from seeing everything from all aspects of the matter and definitely does not resolve it from a practical standpoint as far as its actual materialization is concerned. But in the distant future it is already apparent through a foggy perspective that it is something we can surely dream about, something captivating and important.

Tsiolkovsky's perception and perspicuity are striking. Meditating on his concepts that were published in 1903, we will even more value his imagination and foresight. (This was two years before Albert Einstein published his paper *On the Electrodynamics of Moving Particles*, regarding his special theory of relativity, in 1905.)

Such tests of the propulsive force of gaseous fuels were beyond the capabilities of the era and were not even started for several decades. Even then, doubts rise if Tsiolkovsky considered or knew that hydrogen and oxygen could be cooled and compressed into liquid form, as used at present. The effort to create a high-velocity vehicle reaching the speed of sound, or breaking the sound barrier, was not started until over a decade later by the Russian physicist Sergei Chaplygin with his studies on gas dynamics.

Tsiolkovsky hoped for positive results on the publication of his theory of jet propulsion applied to space travel, but little came of it, as with previous publications. There were no responses in technical reviews, nor was there wide distribution in scientific circles in Russian (and none at all abroad). It was ignored for the most part. Tsiolkovsky was so far ahead of his time that his concepts were beyond the technical faculties of his readers, even those of the Academy of Sciences, and they could not fathom any importance such notions could have for the present. Tsiolkovsky went into a depression, but shortly after, he went on with his work.

With Tsiolkovsky's preliminary analyses of the vertical and inclined ascent of the rocket, its landing with its return to Earth, and the role of gravitational force, he outlined a plan for further research. Tsiolkovsky well understood that his work was just the first step in breaching a grandiose frontier that was to include all the problems and negative aspects associated with his propositions and positive features: the length of time a traveler could exist in outer space, loss or exhaustion of fuel, aerodynamic heating of the surface of a vehicle during acceleration, duration in outer space, and such questions. Tsiolkovsky collected all such problems that eventually needed resolution for a practical space flight and wrote a subsequent treatise with the intention of publishing them in one of the next issues of *Scientific Review*, but then Filippov passed away and the copy that was submitted for publication immediately disappeared, without a trace, along with other papers that Tsiolkovsky had lent to Filippov for his opinion.

This more than just agitated Tsiolkovsky, who went to Moscow to ascertain what occurred. He discovered that ten copies of the next issue had been printed with his article, but they were confiscated by some unknown party; and so nothing remained for Tsiolkovsky to salvage, except his original manuscript at home. This treatise never did find its way into print. As a result Tsiolkovsky withdrew from the public scene.

In 1905, Tsiolkovsky and wife Varvara found a new residence at the end of town near the Oka River. It was a distance from the school where he taught, but with his bicycle the difference in time was not great. The residence included a shed and barn that Tsiolkovsky converted into his workshop and library, and which he coveted for the balance of his life.

Matters at the new home went well until the spring of 1908, when the Oka River overflowed its banks and flooded the city. Again as with the flood in Borovsk, all that the family possessed was ruined, including all of Tsiolkovsky's manuscripts, equipment and instruments.

The previous owner extended his assistance and the entire residence was remodeled, and the family was happy again.

For all practical purposes Tsiolkovsky isolated himself from the public scene shortly after his disappointment with the publication of *Exploration of Space Travel* in 1903. For the next seven years, he concentrated on his family and teaching. Indeed, the impact of his son's suicide forced Tsiolkovsky to concentrate on his family and domestic needs rather than promoting theories regarding space travel. He was essentially forgotten by the Academy of Sciences during this interval. Only one short treatise was published and this was in 1910, titled *Jet Propulsion as the Engine in Flight in Vacuum and Atmosphere*. Otherwise there was no notice of Tsiolkovsky in any journal or discussion.

In one of his autobiographies, Tsiolkovsky mentions that he felt his entire life to be guided or regulated by higher powers. There was some pre-ordained destiny for his life and which he had to accept. During this interval he studied the Gospels and concluded that Jesus Christ was the greatest of moral instructors, although he did not recognize him as being divine.

It was not until summer of 1911 that Tsiolkovsky started to again circulate with other parties interested in flying machines, whether aerostats, airplanes or rockets. He was invited to the Second Air Travel Congress by a new associate, Nikolai Egorovich Zhukovsky, but he did not attend.

A family incident occurred at this time that did not bring them any benefit. In September 1911, their daughter Lubov Konstantinovna was arrested by the police and their home was searched. They found an agenda published by the Russian Social Democratic Labor Party and literature written by Karl Marx, Friedrich Engels and Vladimir Lenin, in her possession. All of this was confiscated and simultaneously the police also confiscated many of Tsiolkovsky's personal notes and letters. Lubov was in custody for a while and then released, but the confiscated documents were never recovered.

In parallel with all his other efforts in promoting his concepts, Tsiolkovsky likewise endeavored to acquire patents. One document from his archives that survives is a personal letter from the USA patent office:

> This privilege is assigned to Konstantin Tsiolkovsky, a resident of the city Kaluga.
>
> The United States of America.
>
> To those to whom this pertains. The present company confirms that Konstantin Tsiolkovsky, from Kaluga, Russia, has delivered a request to acquire a patent on his new and original invention of a dirigible shell, an aerostat. After considerable research on the subject the patent office has determined that he will received a patent as protected by law, dated November 21, 1911.
>
> Note that Konstantin Tsiolkovsky, a resident of the city Kaluga, Russia, a professor of physics and mathematics, informed us that he invented a shell that is impermeable to gas for use in airships.

Subsequent to this, Russia, England, France, Germany, Norway, Belgium, Italy and Austria officially confirmed Tsiolkovsky as having the supremacy to this concept and presented him with the full right of acquiring all income from its use.

Tsiolkovsky hoped these countries would utilize his concept, but all of them essentially ignored it. Their attitude was one of indifference, just as in Russia. As with all the other disappointments, Tsiolkovsky managed to live through this one also.

At the same time, even with the success he gained, Tsiolkovsky expected too much from the general population and scientific community with his research and future intentions. He published a brochure titled *A Simple Project of a Completely Metal-Clad Aerostat Manufactured from Corrugated Steel*. But there was no response from the Academy of Sciences and again his research stagnated.

Subsequently Tsiolkovsky advertised the following, hoping young people who were intrigued with space exploration would be willing to join him as a team:

> Those interested in jet propulsion for outer space travel and having the desire to be some type of participant in my efforts, to continue my work, to evaluate it and in general move it forward one way or another, must first learn all that I have to offer, as difficult as it may be. But since I only have one copy of my *Exploration of Outer Space*, I want to publish it in a complete form and with a supplement, and assistance is needed.
>
> Those desiring to participate in this work must provide me their address. If I acquire a sufficient number of persons, then I will publish the new book and each one will cost no more than a ruble (about 6 to 7 pages long, consisting of about 100 lines of text).
>
> I warn you that this edition is a very serious research composition and will contain a mass of formulas, calculations and tables.
>
> I relate to you my address, to those who want to participate in my efforts: Kaluga, Korovinsky Street #61, K.E. Tsiolkovsky.

But there were no respondents. Later Tsiolkovsky reprinted the advertisement in a Moscow newspaper and a few people responded. Even with no financial assistance from them, Tsiolkovsky proceeded to publish a supplementary brochure where he expounded on his five theorems of rocketry. He also indicated the difficulty of manufacturing a rocket that would serve its purpose, including fuel tanks, jet propulsion discharge, passenger compartments, storage, steerage and navigation. Tsiolkovsky also de-emphasized the possible use of atomic energy for propulsion, since little to no research was yet accomplished on this topic.

Tsiolkovsky was greeted with only a lukewarm reaction with this latest brochure. One Moscow newspaper published the following statement:

> With a precise calculation the author proves that oxyhydrogen gas and even gasoline can successfully replace radium. He states it is possible to acquire materials that are able to tolerate the extreme temperatures of burning gases and not melt. Some of these materials are already known, such as tungsten and carbon. A search proceeds for more applicable substances for the manufacture of the firing and exhaust duct. Using refractory material as a liner will also help keep the metal ducts cool. But this is a task for future experiments and research.

With such a dearth of enthusiasm now in both arenas—aerostats and rocketry—Tsiolkovsky was ready to again surrender as he did in 1903. However, once more he attempted to advertise his concept of the metal-clad steerable dirigible, this time abroad. He published another small brochure, *Defense of the Aerostat*, where he announced having acquired a patent on his metal-clad dirigible and its recognition by several countries, and then he advertised for assistance in constructing a larger model for wind tunnel tests:

> I propose to individuals and communities to build for experimental purposes a metal-clad shell of a medium size. I am ready to provide all cooperation. I have a model that is about 7 feet long. But this is too small. In the event of a successful functional medium size unit, I am willing to yield to my investors one or some of my patents.
>
> If I had the financial means, I would myself construct a medium size model that would function and then sell it. Even then, if someone is willing to purchase my patents from me, I would be willing to share 25% of my profits with him, and use the balance for fabrication.

As with Tsiolkovsky's other attempts at acquiring assistance, he had no respondents. Then he put out a second brochure with the invitation,

> Please come to my home and inspect any of my models, any day of the week, at 6 o'clock in the evening.

As with the previous advertisement, no one indicated any interest.

Tsiolkovsky did receive a couple of awards worth noting. In August 1906, he received the Order of St. Stanislav, 3rd Degree, for conscientious efforts in the area of science and technology. In May 1911, he received the Order of St. Anne, 3rd Degree, for his effort at the Kaluga school where he taught.

So seven years passed with Tsiolkovsky in a state of oblivion to the scientific and academic world, but with his short treatise published in 1910, his name surfaced and the editor of *Air Flight Messenger*, B.N. Vorobyev, asked him to submit another article. Tsiolkovsky wrote his response to the editor:

> I reworked some facets of my question on the ascent of a jet-propelled vehicle such as a rocket to outer space, my mathematical conclusions based on scientific data and which were checked many times. All of this points to the possibility to ascend to heavenly regions with the help of such equipment and perhaps to found a settlement beyond the regions of Earth's atmosphere.

Tsiolkovsky proposed to Vorobyev a second volume to his *Exploration of Outer Space*, but one that would be serialized over several issues, each issue advertising the contents of the next to keep up readers' attention. The editor responded positively, except for noting that the readers would need to realize that these were theories generated by Tsiolkovsky and that he still did not have sufficient scientific basis to put them into practice. It had to be presented in more of a science fiction format than plans for actual space travel.

The publications started with issue 12 in 1911 and continued to issue 9 in 1912, but in Tsiolkovsky's view much was lacking in these articles. This was still an unknown sphere of knowledge that Tsiolkovsky was communicating to his readers, and much of it was invisible to them, such as gravity or the vacuum state of outer space, or what was completely unheard-of, such as jet propulsion. Overcoming gravitational pull seemed to be a war with an intangible enemy and the manner of victory was to escape by ascension. Tsiolkovsky described gravity as a wall around Earth that had to be broken by means of a high velocity penetration and so allow Earth's residents to travel to other planets and regions of outer space.

Responding to questions from a person who read several of the articles, Tsiolkovsky made the following comment about the composition of spatial bodies:

> There was a time and not very long ago when the idea of the possibility of knowing the composition of heavenly bodies was considered insane, even among the famous scholars and thinkers of the era. Now this time has passed. The thought of the possibility of a much closer and unimpeded study of the universe, I think, seems to be more fundamental at the present time. To place your foot on the soil of asteroids, to handle a rock on the surface of the moon, to build a moving station in the ether of outer space, to form a circle of residences rotating around Earth or the moon or the sun, to observe Mars from a distance of a few dozen miles, to descend on its satellite or even on its surface—all of this perhaps is mind-boggling. However at the moment of the application of jet propulsion a new great era in astronomy begins, the epoch of greater concentration on the study of heaven.

Tsiolkovsky's series of articles was a tremendous success. There was no doubt that he was motivating the minds of his readers to start looking beyond themselves and into the surrounding skies, with its planets and stars and galaxies, to wonder what was up there and if it could be reached. In later years Vorobyev reflected on the impact of Tsiolkovsky's articles:

> The resonance that occurred was massive. Scientific and technological and popular magazines all responded with comments, and regular newspapers and inventors. A large number of authors started to promote projects dealing with constructions of jet propelled aircraft popularizing Tsiolkovsky's ideas of the possibility of defeating outer space by escaping the atmospheric regions of Earth and traveling

> further, and fantasizing about communicating with super-human societies of distant galaxies.

The reason for Tsiolkovsky's popularity in 1911, as opposed to 1903, is simple to explain. Earlier, the topic of flying, much less space travel, had only a few enthusiasts, but seven years later it had succeeded in growing into a very popular topic. Tsiolkovsky's book *Exploration of Outer Space Using Jet-Propelled Rockets* was reprinted also in 1912, and it brought him a different type of wealth: a circle of friends who valued his ideas—and this made his popularity increase all the more. Some of these men were Vladimir Vladimirovich Ruimin, a Russian engineer and science instructor, and Yakov Isidorovich Perelman, author of many Russian science books, and Professor Nikolai Alekseevich Rienin, himself involved in the fervent propagation of cosmic ideas. These new friends supported Tsiolkovsky in later years during difficult times. Ruimin published an article complimenting Tsiolkovsky in the magazine *Nature and People* in the September 1912 issue, which included the passage below:

> I read it once and then I read it a second time with a pencil in hand, ascertaining all the mathematical computations of the author. Yes! What a concept! Tsiolkovsky achieved a conquest over the elements of space. This is genius disclosing to future generations a route to the stars. His ideas need to become accessible to larger masses of readers. And perhaps among them we will find people who will penetrate the greatness of Tsiolkovsky's thinking and will be able to help him draw closer to its materialization. They only need to prove that he is right, that flight in the vacuum regions to other planets of outer space is actually possible. But this is not a simple scientific fantasy, but a very real proposition.

Ruimin supplemented his acclaim of Tsiolkovsky with an explanation to the magazine's readers of the mechanism of jet propulsion in a rocket and why the rocket would be the sole means of penetrating the cosmos. Ruimin also included the possible use of atomic energy in future cosmic space stations. This was likewise an idea ahead of its time.

A new imposter resurrected at this time, attempting to exploit Tsiolkovsky's achievements. As with David Schwartz some 15 years earlier, a person in France came along. He was Rober Esno Peltri, a young electrical engineer who also had some background in aviation and who visited Petersburg in early 1912. He claimed to have made these discoveries before Tsiolkovsky, although he was a fraud As Tsiolkovsky related in his memoirs:

> After I published the continuation of my work in a widely distributed and specialized journal, then immediately in France an

impressive and intimidating person announced that he had created these same rocket theories earlier.

The second part of *Exploration of Outer Space* was published in May 1912. In September 1912, Ruimin's article mentioned above was published, and this received a wide circulation and popularity. But it was not until a meeting of the French Physics Society, in November 1912, that Peltri announced his discoveries to the group of physicists and mathematicians gathered. Prior to this time Peltri had not published any articles on the topic, except for the possible use of atomic energy in the future as a motive force. The society's members reprimanded Peltri, accusing him of insulting their intelligence. Peltri's conclusions contained too many similarities to Tsiolkovsky's articles, and Peltri's presentation was definitely amateurish. The close proximity of Tsiolkovsky's publications and Peltri's announcement likewise caused suspicion, since many of Russia's scientists also spoke French and were in communication with French scientists.

Others, however, lauded Peltri since he convinced them that an atomic energy motor could be used to supply motive force; plus, he was French. Some years earlier the French Society of Civil Engineers had announced a reward for an original design of a lightweight engine or motive force and Peltri obviously was eager for the prize. Following Peltri's visit to Petersburg to meet with Russian physicists, it was impossible for him not to know of Tsiolkovsky's research on the subject.

The following year, on November 20, 1913, Perelman presented a report to the French Society of Physicists echoing the voice of the Russian scientific community. He explained Tsiolkovsky's concepts of the utilization of jet propulsion on rockets intended for space travel, his presentation being a combination of the results of Tsiolkovsky's experiments and scientific data and his science fiction compositions. The intention was to convince the society of the real viability of such ventures in some future era when manufacturing could produce such vehicles and equipment, and that Tsiolkovsky's concepts were very promising. Indirectly, of course, Perelman erased any aura of validity surrounding Peltri's presentation, claiming that Peltri had plagiarized Tsiolkovsky's research with copies of his treatises that were translated into French and sent abroad by Russian scientists. On his return to Russia, Perelman wrote an article repeating much of his French presentation for Russian readers.

Tsiolkovsky, having read the article, wrote to Perelman:

> Deeply respected Yakov Isidorovich. I received your letter and the copy of the article in the *Contemporary Word* and read them enthusiastically. You and Ruimin have resolved this conflict I incurred and I do not know how to thank you. Nonetheless, the result of the

> matter compels me to continue in rocket research and discover even more that is new.

The matter did not end here, as in late 1914, a Russian scientist, K.E. Veigelin, published an article contradicting the propositions made by Perelman and repeating the presentation of Peltri, but public protest rose against Veigelin, informing him of Tsiolkovsky's earlier research. Eventually Peltri desisted in his ambitious efforts for the next decade, but in 1927 made another presentation to the French Astronomical Society, which was not heartily accepted. It was not until 1928 that Peltri again published some research on the subject, and he finally gave credit to Tsiolkovsky in his acknowledgements.

The following year Tsiolkovsky again attempted to gather some enthusiasts at the school auditorium where he taught. In the April 1914 issue of *Kaluzhski Courier*, the following was published:

> The first public exhibition of our popular inventor, K.E. Tsiolkovsky, wanting to familiarize his denizens with the results of his work of many years in the area of the construction of a metal-clad steerable aerostat, occurred on March 27 at the Shakhmagonov School auditorium, a room that can only accommodate a few people. Not including the few who are members of a local nature society, which was the group that arranged the lecture, about 20 or 30 adult students of the school attended with about a handful of others.

In reality, the exhibit was a total failure, but Tsiolkovsky's aspirations were still not dimmed.

That March he received an invitation to Petersburg to attend the Third Air-Flight Congress, and the Academy promised him 50 rubles to cover expenses. Of course, Tsiolkovsky agreed to attend, and he decided to bring with him a couple of aerostat models to exhibit and to listen to the opinion of those specialists in the subject of aeronautics. Together with Tsiolkovsky was his assistant P.P. Kanning.

The participants of the congress listened to Tsiolkovsky's report—he did not read it himself, due to a sore throat, but Kanning did. The report was accompanied by a demonstration. A handmade pump forced air into the deflated shell indicating how it would expand and change shape as pressure was applied to it.

As with Tsiolkovsky's many other attempts at selling his concept, his collapsible metal-clad aerostat did not capture the attention of the audience, on which Tsiolkovsky so depended in order to continue his work. There was just no interest in it. What did catch their attention were the exhibits and presentations of heavier-than-air aircraft. (The Wright brothers were already traveling in Europe in 1909, selling their design.) Tsiolkovsky and

Kanning packed their bags and boxed their model aerostats and hurriedly returned to Kaluga again disappointed. Some of Tsiolkovsky's earlier supporters abandoned him at this congress.

On August 31, 1914, Germany declared war on Russia.

Tsiolkovsky was now 61 years of age; his health was failing and he needed to start considering what would occur to him and what would become of his family in later years. He had no savings due to expending all of his income, every ruble that he earned, on his research and experiments and on supporting his wife and younger children. He decided to turn to the Ryazan Provincial Government with a claim based on his nobility, along with his auditory disability, in order to qualify for some property that had been allotted to his family in earlier generations. The property that Tsiolkovsky had his eye on was in Chernomorsky Province at the east end of the Black Sea in southern Russia. He sent a letter with this request on December 21, 1914. But the Russian government responded that land was scarce and they were in no position to assign him a parcel. World War 1 was also in progress and most of the country was militarized.

In 1914, Tsiolkovsky succeeded in publishing another thin little brochure, *The Second Law of Thermodynamics.* The brochure was thin with nothing appealing on its cover, but it was evidently printed and distributed without a clear approval of the government censurer. Inside of it was a bold summons to deliberation. Tsiolkovski refuted the premise of an end of the world, an opinion prevalent in the Russian Orthodox church.

Tsiolkovsky's theory was originally compiled in 1905, when he communicated it to a fellow physicist, Orest Khvolson, at the Academy of Sciences in Petersburg, but it went no further. It was too philosophically inclined for him to consider seriously. Tsiolkovsky then essentially ignored the topic due to more pressing items on his agenda with his aerostat and rocket designs.

Tsiolkovsky's attempt was to reinterpret from his own standpoint the second law of thermodynamics and magnify it in a yet unprecedented manner. This brochure brought people's attention to another Tsiolkovsky, not the investigative researcher or "mad scientist" that they knew, but a serious philosopher delving into ontology and materialism. He knew his interpretation would conflict with the established doctrines of Orthodoxy as far as the universe having a specific start and termination, so he published his brochure without their review since he knew it would not be approved. His approach coincided with materialist doctrine that everything consists of matter including what is labeled spirit, as opposed to the religious view of a spirit or non-material realm. The philosophical source of Tsiolkovsky's treatise was Friedrich Engels, to which Tsiolkovsky then applied his results acquired from his study of atoms as the fundamental building block of matter and the course of the universe being reiterative.

> So to agree with the zealous conclusions of [Rudolf] Clausius and [Hans Peter] Thomsen, the temperature of bodies strives for balance, toward one defined mean temperature, in other words, the entropy of the universe uninterruptedly grows. This means the time will arrive when suns will exhaust and extinguish themselves, the worlds will cool and animal life will terminate.[1]

> But this will not occur if Clausius' postulate is not recognized as a rule or law. The universe has existed from long age, and it is difficult to imagine that at one time it did not exist. But if it already exists an infinite time, then it should have entered a state of temperature equilibrium, the extinction of the sun as a heat source, and a general death. But since this has not occurred, then there is no such law, but what we have are periodical iterative incidents.

> First is thought and then action. Without thought—action cannot start. It is good if this article will awaken the thinking of young mind and compel them to perform experiments similar to mine. Personally I do not see the need to further perform experiments and in part due to it being all the same to me, since they did not believe in experiments in regard to the resistance of air.

This thread of philosophical thought of the existence of the universe from infinitely in the past and infinitely into the future, and the iterative birth and regeneration of suns and solar systems and galaxies, and clumps of galaxies (to which he gave the appellation of astronomical units) now pervades all of Tsiolkovsky's future writings. From 1919 on, he was as much a philosopher as he was a scientist, and most of his compositions during the latter 15 years of his life had a dialectical materialist vein in them, supplemented as needed with his scientific discoveries.

During the years of the First World War and Russian Civil War, between 1914 and 1919, Tsiolkovsky again isolated himself in Kaluga, primarily to protect himself from the impact of both events. Viewing Tsiolkovsky from another side, due to his naïveté in regard to political issues and the militarist forces dominating Russia, it was difficult for him once he became cognizant of what was occurring. Tsiolkovsky was so enveloped in his scientific experimentation and astrophysical research that he did not notice the shift in the political environment of Russia. In his memoirs he blamed that on the gift of his genius, as it caused him to be impervious to surrounding events and caused him much sorrow as a result.

[1] Tsiolkovsky calls this a warm death in other composition.

In 1916, Tsiolkovsky published his short treatise *Sorrow and Genius*. He hoped to encourage himself by reflecting on the failure of the state to support and take advantage of the benefit that geniuses could offer, including himself. This failure caused universal suffering. If a genius surfaces in some country at some time, the state has the obligation to promote him and utilize his talents to the maximum amount reasonable possible, and the genius will likewise not suffer any sorrow.

Deprivation, suffering, misfortune, abandonment of the aged, destitution with an abundance of natural resources available, homelessness and unemployment, hunger, and other social disasters, occur as a result of the state failing to heed the genius. As Tsiolkovsky wrote:

> If geniuses were searched for and discovered, then the most horrible of misfortunes and sorrows, those which even seem to us at the present to be inescapable, would be estranged. Geniuses performed and do perform miracles. Who does not know this?
>
> The genius provides health and prolongs life. Chemistry livens the dying, heals the handicapped, gives vision, audio and on and on. Technology allows a human to be stronger than a tiger, faster than a deer. He gives the human wings. He teaches the people to construct, and subjects nature as a slave to the human's benefit.

This article was the first of Tsiolkovsky's series of philosophical treatises and it brought another side of him to the attention of readers: his compelling view that science can be used to resolve all the problems that humanity needs to face. It was Tsiolkovsky's drive to use science in a practical manner to improve society. In general, Tsiolkovsky welcomed the Russian revolution, since he felt the new idea for society was based more on science and less on superstition, and the socialist agenda agreed with his.

In 1919, Tsiolkovsky published his brochure, *The Kinetic Theory of Light*. In one respect it appended his *Second Law of Thermodynamics*, published five years earlier, and again he relates his conviction that if the theory of the warm death of the universe were plausible, then our eyes would now be witnessing the gloomy scene of an extinguishing world. Over the course of many years, Tsiolkovsky defended his idea of "the universe's eternal adolescence."

Even if the Academy of Sciences under the Imperial Russian government did not provide Tsiolkovsky the support he needed or expected, whether financial or moral, the new Soviet government did. In light of a combination of Tsiolkovsky's aerodynamic efforts with regard to aircraft of all types—aerostats, rockets and airplanes—along with his treatises including the philosophy of dialectical materialism, the Soviet Academy of Social Sciences on July 13, 1918, extended him an invitation to be part of the construction of the Soviet Union. Immediately Tsiolkovsky informed the Academy that his ideals of social reconstruction were close to what was contained in the new Soviet Constitution. He also asked for financial support, without which he could not continue. At the time Tsiolkovsky only received an income of 35 rubles a month from teaching.

On August 26, 1918, the Soviet Academy elected Tsiolkovsky as a member. The following is the text of the official letter sent to Kaluga:

> The Soviet Academy cannot rectify the mistakes of the past, but it will strive at least for the near future to display the necessary cooperation to your sincere aspiration to manufacture something useful for people. Disregarding the extreme deprivations, your spirit is not broken. You are not aged. We still expect much more from you and we want to remove from your life any material obstacles, anything that interferes with the total success and completion of all objectives within the capabilities of your genius.

The Academy proposed to Tsiolkovsky that he relocate to Moscow. He could work at a local collective farm, and the doors to all the libraries would be open to him. But it was too late. By this time in Tsiolkovsky's life, he was aged—now 61 years of age—and ill, and weak due to a lifetime of devotion and work and worry under the most destitute of conditions. He did not have any strength to move to Moscow. So Tsiolkovsky requested that he be allowed to work in Kaluga. Moscow was insistent on his relocation and even offered to find work for his daughter and subsidize her if necessary. Tsiolkovsky again had to deny the request and wrote in return:

> My deafness is difficult for me and intolerable to others, also my agedness, illness, I am gaunt from hunger, and I still have four children to raise with only the help of my wife. Moving to Moscow will destroy me. Here I can be calm and somewhat content with your support. Under the conditions here at my home it will be easier for me to deal with extreme circumstances and perform the work you require.

Tsiolkovsky remained in Kaluga. The Academy approved a subsidy consisting of 835 rubles every two weeks, a substantial sum at the time, to begin June 5, 1919. As a result of the decision, Tsiolkovsky immediately quit his teaching position. To repay the Soviet government for their benevolence, Tsiolkovsky decided to exert effort on behalf of the Red Army in the south of Russia. His intent was to build a dirigible for use to transport troops and supplies. But the concept to replace their existing railway transport system with a dirigible was beyond their ability to fathom and the matter terminated quickly.

All of a sudden Tsiolkovsky was full of energy and a desire to serve his country. He was no longer the recluse or mad scientist of earlier decades, as others had viewed him. The new era of Soviet Russian Socialism motivated Tsiolkovsky to apply himself to the general welfare. He also composed a series of articles on the operation of the Soviet state in regards to health care, economy, education, property rights, marriage, family, and other topics, all to the benefit of the individual and the state.

However, the situation changed. Other scientists in Moscow who were not enthused with Tsiolkovsky's earlier work in aerodynamics and rocketry maneuvered the Academy to suspend his membership (also on the grounds that Tsiolkovsky refused to move to Moscow). The motion was led by a revolutionary and scholar named Morozov, who little appreciated Tsiolkovsky's efforts. As a result, his subsidy from the Academy was terminated. It was the next tragic event in his history. After leaving his teaching position, Tsiolkovsky resumed that vocation again on November 1, 1919, at the Kaluga Children's School, teaching 6th grade. He had no other choice.

On October 5, 1919, Tsiolkovsky's son Ivan died of malnutrition and heavy labor. One more tragedy for him and the family.

For an aged and ill person, the teaching work was difficult. The classrooms were not heated and half the time the room had no lighting. The termination of Tsiolkovsky's pension still did not terminate his financial support from individuals still faithful to his cause. V.V. Assonov, son of the above mentioned publisher V.I. Assonov, published several of Tsiolkovsky's subsequent articles: *Transport by Air, The Gondola of a Metal-Clad Dirigible, The Wealth of the Universe,* and *the Kinetic Theory of Light*. He was undefeatable in his struggle on behalf of his dirigible design, a fervent opponent to the theory of a warm universal death, and continued to preach his thoughts of the universe's eternal adolescence. In addition he received help indirectly from Perelman who wanted to republish Tsiolkovsky's science fiction book *Beyond Earth*, in 1918, but only published a few initial chapters in a local magazine and was unable to finish it due to the unavailability of paper during wartime. But then Perelman made arrangement for a Kaluga society who valued Tsiolkovsky's efforts to publish it in August 1920, although they were only able to print 300 copies. Copies nevertheless circulated and the book was reprinted in both Berlin and Vienna by Russian publishing houses. It is Tsiolkovsky's longest single composition, consisting of over 70,000 words. V.V. Assonov wrote the prologue to the book:

> It seems to us that enough has been said to turn our attention to Tsiolkovsky's present science fiction narrative, where essentially there is less fiction and more science in its text, and the context always has firm scientific data to support it. The book is the product of very meticulous and difficult mathematical research.

Tsiolkovsky placed the events of the book into the future of the year 2017, exactly 100 years after the Russian Revolution. All of Russia seemed to be under siege with the war against Germany and the subsequent civil war between the Red Army and the White Army, and it likewise affected Kaluga, with the reports of battles, the dearth of commodities available in stores, and the regular announcements of the casualties of war. Tsiolkovsky, who was in earlier years opposed to war, dreamed of peace in Russia and attempted in one section in the book to portray specifically Russia as well as the world as abiding in a general state of peace and contentment. He wrote in *Beyond Earth*:

> There was one government over the entire Earth—a congress consisting of selected representative of all the countries. It had already been in existence over 70 years and resolved all disputes pertaining to humanity. Wars were impossible to conduct. Misunderstandings between nations were reconciled through peaceful means. The army was initially very limited in its application and soon the army became a labor force. Under such prosperous conditions the population of Earth over the previous 100 years tripled. Commerce, technology, culture and agriculture attained a significant success.

In one respect, the above passage is prophetic. If the events in Tsiolkovsky's book occurred in the year 2017, then 70 earlier would be 1947 when this organization already existed, and the United Nations was formed in 1945.

As much faith as Tsiolkovsky had in the future, and especially socialism, he was not immune to suspicion from the Bolsheviks during the years of the Russian civil war. On November 17, 1919, Tsiolkovsky was suspected of collaborating with the White Army and subsequently arrested and interrogated. He was released after a couple weeks in custody. Information on this incident is meager.

Apparently Tsiolkovsky had been communicating with a man named A.J. Feodorov, from Kiev, who was also an official in Anton Denikin's White Army. He corresponded with Tsiolkovsky regarding construction and use of dirigibles by the White Army against the Red Army, but Tsiolkovsky refused to discuss the issue. Feodorov came to Tsiolkovsky's home to visit him and get some information about dirigibles, but nothing occurred and Feodorov was told to leave. Later the Soviet KGB came to the house and arrested Tsiolkovsky; they took him to Lubyanka Prison in Moscow. He described the incident in a letter to his friend Vishnev dated May 4, 1920.

> This is what happened. For a long while I corresponded with an aircraft pilot in Kiev named Feodorov. I personally did not know him and I never even saw of photograph of him, but he indicated a great

> interest in my aerostat. Then he wrote to some 3rd person, unwarranted and without any basis to do such a thing, and told them I was familiar with the activities on the Eastern front. This letter somehow ended up in Moscow in the Offices of the Committee for State Security (KGB). Of course even though nothing was found since there was nothing to find, I was nevertheless arrested and brought to Moscow. After two weeks, and thanks to your intervention, they took notice of me and had no other choice except to acquit me. The interrogating official seemed to be a good person and treated me respectfully.

No specific charges were brought against Tsiolkovsky and neither was the specific reason for his arrest even documented, and any documents dealing with the arrest vanished with his release. Tsiolkovsky was released; he found himself on the street in front of Lubyanka prison without money and hungry. He returned home in an empty railway car. On the journey home he injured his leg. After a few days Tsiolkovsky regained his health and he put the incident behind him.

Tsiolkovsky desisted from further communication with Feodorov, but he then accused Tsiolkovsky of sabotage and defamed him and threatened him with another arrest. But nothing further resulted from the matter.

In June 1921, the Kaluga Society for the Study of Nature brought the matter of Tsiolkovsky to the Scientific-Technological Society of Kiev to reconsider a pension for Tsiolkovsky under the conditions of his destitution and illness. He was now 64. The request was transferred to Petrograd, and a historian of Russian aviation, A.A. Rodnikh, and a professor of aeronautics, N.A. Rienin, mentioned above, became involved. They invited Tsiolkovsky to relocate to Petrograd and become a physics or mathematics teacher at an institute. Of course this was difficult for Tsiolkovsky, as with the other previous opportunities to relocate to either Moscow or Petrograd.

Persistent intervention by friends brought some results as far as re-instituting Tsiolkovsky's pension, which he had lost the previous year. On October 1, 1921, the Ministry of Education of the USSR elected to double Tsiolkovsky's pension. On November 9, 1921, the final amount decided was 500,000 rubles per month, but still less than the previous amount in terms of new rubles. (The ruble was readjusted in value on January 1, 1922, to 10,000 old rubles to one new ruble.)

Finally Tsiolkovsky's hardships came to an end. He wrote his second letter of resignation to the school where he taught:

> I have reached the age of 64 years, I suffer with chronic bronchitis, my intestinal tract causes me regular difficulties, and my deafness and general weakness force me to resign from teaching. I request you to consider me no longer employed at my civil service job beginning November 1, 1921.

Now there was nothing in the way to interfere with Tsiolkovsky completely dedicating himself to his research. He happily wrote to Perelman:

> I quit the school. Due to my age and health this was too excessive for me. Now I can devote myself more intensively to my beloved work—jet propulsion equipment.

However little new was actually discovered by Tsiolkovsky from this point to the end of his life, almost 15 years later. His later articles on technology only refined his previous work. Most of his compositions were his cosmic philosophy, the possibility of space travel, the purpose and cause of the universe, the reason of existence, defining why the universe continues infinitely, or else the compositions were of a socialist nature with the application of his scientific observations. Included were also a couple of autobiographies.

In April 1924, 25 members of the N.E. Zhukovsky Air Force Union Academy gathered for a discussion on interplanetary communications. At this first session they decided to invite Tsiolkovsky to provide them some scientific guidance. The secretary of the Academy was M.G. Leiteizen, and a regular correspondence began between him and Tsiolkovsky at this time. Tsiolkovsky first sent Leiteizen a copy of his various articles on the concept of flight beyond Earth's atmosphere. This new group was serious about their intention to send a rocket into space, and to the altitude of 70 miles, using jet propulsion. They insisted on Tsiolkovsky's presence, but alas, as with all the other invitations, Tsiolkovsky had to decline due to illness. Instead, he wrote out his lecture and it was recited at the Academy meeting by Professor Lapirov-Skoblo. The following day Leiteizen wrote to Tsiolkovsky about the results of the meeting:

> Most respected Tsiolkovsky.
>
> Yesterday evening's meeting was dedicated to interplanetary excursions and it was concluded with exceptional success. Tickets were all sold for the event long before the lecture, and the administrator of the museum was forced to invite some policemen in order to restrain others from entering. The literature that we had available, and most of it was Perelman's, was immediately sold out. What is sad is not having your literature available for sales.

The lecture apparently did impress a lot of people. One of the attendees, G. Kramarov, recorded that about 200 attended. This meeting was the first of its type in the world, but regrettably the Union Academy membership did not last much longer; it was closed. However, the excitement caused by the meeting was enough to encourage Tsiolkovsky to continue his work and it motivated him to travel to Moscow for a subsequent meeting of members of the Association of Natural Philosophers, which was headed by a person who was now very devoted to Tsiolkovsky, Yakov Yazikovich Rappoport. However, some who were totally against the concept of rocketry in space travel were also to attend this session.

When the railway car arrived at the station in Moscow, an aged gentleman stepped out of the car and onto the platform. An immense crowd of devotees gathered to great him and they drove to the auditorium where Tsiolkovsky was to deliver his address. A.G. Stoletov and N.E. Zhukovsky and other physicists accompanied him.

From the testimony of witnesses who attended, the conclusion of Tsiolkovsky's presentation ended with a storm of applause. Press photographers surrounded him and the general response of the meeting was to incorporate Tsiolkovsky's concepts in their future rocket designs.

Tsiolkovsky was also invited to a discussion on a trans-arctic flight to be held on July 22, 1924. The means of travel was the dirigible, now with attempts planned by Roald Amundsen, a Norwegian. (Such a flight was accomplished in May 1926, by dirigible, but after the failure of his flying boats in 1925.) Now the USSR wanted to proceed with the matter. However, due to poor health Tsiolkovsky did not attend.

In 1926, Tsiolkovsky again published his *Exploration of Outer Space Using Jet-Propelled Rockets*. However, this time it included considerable new research material and information. He invited readers for a discussion on a practical means of implementing his designs for a production model. As before, some started right at the beginning with ideas that had long since been discarded: a long gun-barrel design to shoot a projectile into outer space; electromagnetic cannons, and others. Tsiolkovsky again promoted his idea of a using a jet-propelled rocket, or perhaps dividing the effort between two rockets: one that would travel just beyond Earth's atmosphere and then a second that would take off from that altitude and continue into outer space.

As with Tsiolkovsky's other meetings, the readers assessed his concepts as being far from reality and any feasible materialization.

In 1927, another newly-formed group, the Moscow Association of Inventors and Designers, chose to celebrate Tsiolkovsky's 70th birthday. They decided to organize an exhibit dealing with interplanetary travels and communications. The exhibit was to include all the Russian scientists available on the subject as well as inviting several from abroad, such as Max Valier, Hermann Oberth, and Walter Hohmann. Of course, Tsiolkovsky did attend and took center stage. A bust of Tsiolkovsky was also erected to his credit.

The exhibit was successful and several thousand visited it over the week it was open in Moscow. At the end of the exhibit, Tsiolkovsky was awarded his bust to take with him to Kaluga. The result was that Tsiolkovsky became legendary, an aura of eccentric awe surrounded him, and the results of his experiments were lauded. He became as large as the universe whose exploration he was promoting, and all his efforts in this direction were finally recognized and the credit due him was now exhibited.

In 1926, Tsiolkovsky proposed a design of a launch pad, dividing the work of the launch between two rockets: the cosmic rocket would be mounted over the ground rocket. Presently we define these as a first stage and a second stage.

Tsiolkovsky's subsequent book, published in 1927, was *Resistance of Air and a High Speed Train*. This was a display of a new means of transport, the use of jet propulsion for a railway system, such that the velocity and force of the jet engine would lift the train up off the rails. Tsiolkovsky's new train would eventually be able to traverse long distances without the need of rails, and so there would be no need of bridges and tunnels. Tsiolkovsky claimed his jet-propelled train could reach a speed of 500 miles an hour. Since the velocity was so high and the train was actually floating over the rails, Tsiolkovsky felt that plastic rails would serve just as well, where needed.

The attempt to test a model failed desperately, since it was impossible to build a small scale jet-propelled railway car with a sufficient compressor for the experiment.

Tsiolkovsky's next proposal was a wheel-less rover, where small jet engines would lift the vehicle above ground just a short amount and then the vehicle would maneuver sideways. Since this design was unfeasible, the next design was an air hovercraft or air-cushion vehicle. This was actually the best of Tsiolkovsky's later designs, since all it needed was an engine, frame and fan. The advocates of this design figured that rivers would be best utilized as the routes for the hovercraft, and it would soon replace automobiles and boats.

Tsiolkovsky likewise realized that his concepts of dirigible design could just as equally be utilized in submarine vehicle design, the difference being the surrounding media: water instead of air. In a 1930 article titled "Dirigibles," he wrote:

> The dirigible can be compared with an underwater vehicle, except that the media surrounding the dirigible is 800 time lighter than water. But it will also be driven by a propeller and will be submerged in water relative to its weight and buoyancy according to the law of Archimedes.

Tsiolkovsky continued with this cosmic philosophy with his subsequent published brochures, "Monism of the Universe," "Cause of the Cosmos," "The Formation of Solar Systems and the Debate over the Cause of the Cosmos,"

"The Future Earth and Humanity," "The Previous Earth, The Present Composition of Earth," "Will of the Universe," "The Unknown Intellectual Powers," and "The Goals of Stellar Travel," all of them filled with suppositions, puzzles, presumptions and conjectures of Tsiolkovsky's desire to penetrate into the realm of the cosmic and corporeal unknown. Tsiolkovsky was also obsessed with the possibility of communication with inhabitants of other worlds, and visiting them. Tsiolkovsky fantasized immensely as he wrote:

> Now in view of the proven possibility of interplanetary communication, we press forward to the unknown with greater attention.

In the brochure *The Previous Earth*, Tsiolkovsky underscored the connection between the development of flora, fauna and the composition of the atmosphere. He noted that the height and composition of the atmosphere changes the mean temperature of the Earth's surface, while the development of plant life and animals changes the composition of the atmosphere, and this is what creates climate. The question of the generation of life always interested Tsiolkovsky. He wrote in his *Cause of the Cosmos*:

> One of the two: either Earth became populated by self-generation or the essence of life was transferred here from other planets. The hypothesis of self-generation is the preferred, since it is the only manner by which life can generate in the cosmos.
>
> We are convinced that mature entities of the universe have the means of migrating from planet to planet, installing life in the balance of planets. I have also proved that the migration of life is possible with the help of the technology of the higher entities, those similar to the human. Then would life have appeared on Earth along with these entities and their high civilization, technical advancement, and facilities of every type. If all of this were to be destroyed by hostile activity, or by some type of catastrophe, for example a massive earthquake, comet, or a fire-ball from outer space,[1] and etc., nevertheless, it is impossible for fossils or remnants of the higher culture not to remain. However, this we do not see. We have found traces of insects and worms that existed in the past, and I am sure we will also discover traces of higher humanity.

A person cannot but be surprised in reading these statements. Of course, Tsiolkovsky never knew that years after his death the hypothesis of invaders from outer space would be promulgated by astronomers and science fiction writers.

[1] The reference here is to the Tunguska Event of June 1908.

Tsiolkovsky's advocacy of the dirigible did not cease. One of his brochures was *Travel by Dirigible*, published July 7, 1935, where he again emphasized that the dirigible is the least expensive method of travel and transport.

Tsiolkovsky's concepts of jet propulsion further progressed with the creation of GIRD on September 15, 1931, the Group for the Study of Reactive Motion. They formed the basic school of the practical design of rocketry in the USSR for the coming decades, headed by seven researchers under the leadership of Sergei Korolyov, who later became head of the Soviet space program. Their motto was, *We were the first to penetrate space.* Tsiolkovsky was only indirectly involved with the group via correspondence with their questions and his answers and advice, and he followed all their events and meetings to the extent he could. He was now 74 years of age.[1]

First on the list of GIRD's tasks was the construction of a jet airplane. The initial working model of a jet engine was completed about a year later in July 1932. The first Soviet aircraft with jet propulsion flew on March 1933, but the engine disintegrated during the test flight. (The Soviets were behind Germany, who flew their first successful jet propelled aircraft in September 1929.) It was not until 1938, that the Soviets had a functional jet propelled aircraft.

The Reactive Science Research Institute made Tsiolkovsky an honorary member of its technical counsel on July 23, 1935. They also designated the letter Ö (Ts) (the first letter in Tsiolkovsky) as the relationship between the initial weight of fuel and the remaining weight of fuel during launch.

On October 9, 1932, in Leningrad, Tsiolkovsky was awarded the Order of the Red Banner of Labor, the highest award to a civilian at the time, for his accomplishments for the USSR and society in general. The award was presented by Mikhail Ivanovich Kaligin.

But it was not only in the area of rocketry that Soviet scientists proceeded. Tsiolkovsky's earliest concepts on a hot-air balloon motivated them to design one that would rise higher in altitude, as high as the stratosphere. On July 26, 1935, a Soviet stratostat reached a successful altitude of over 11 miles. Tsiolkovsky was also given the credit for its design and feasibility.

As with many scientists in the past and present, an inherent fear resides in them that their inventions and discoveries could conceivably be used for war and the military. So with Tsiolkovsky. In 1919, he wrote in his treatise, *Regulations of Society*:

> In reality, militarism eternally threatens every person by forcing him to be subjected to violence and be killed or at least maimed.

[1] The first successful liquid-fueled rocket launch was in March 1926 in the USA by Robert Goddard.

> Let those who want to wage war do so. But if a society forces its members to be part of a military campaign, then it violates the right to life. In any case, those involved in a fight or war must be subject to trial for their crimes.

It worried Tsiolkovsky that his designs for rockets and jet aircraft would be utilized in the Soviet Air Force. Although he was too aged at the time to become involved in this, and even though it would not happen during his lifetime, correspondence between himself and others toward the end of his life indicates his disapproval that his inventions and discoveries should be used for military purposes.

In March 1935, Tsiolkovsky consulted a family doctor in Kaluga, N.I. Sirotkin, about internal pain. Sirotkin diagnosed his condition as a malignant tumor in the abdominal cavity. The doctor recommended he go to Moscow for surgery and further treatment, but Tsiolkovsky refused.

Tsiolkovsky's final article was composed on August 29, 1935, for a local newspaper, and his health declined noticeably from that time. He finally agreed on surgery, but in a local Kaluga hospital. Tsiolkovsky underwent surgery September 8 to remove part of his intestine and the malignant tumor. Tsiolkovsky did not long survive; he only lived 11 days more.

However, a few days after the surgery, on September 13, Tsiolkovsky dictated a letter to Marshal Joseph Stalin and the Central Committee of the USSR, bequeathing all of his work to the people of Soviet Russia.

> Before the revolution, my dream could not be materialized. However October[1] brought a recognition of my self-educated efforts. It was only Soviet authority and the political party of Lenin and Stalin that displayed effective assistance. I felt the love of our national population and this provided me strength to continue the work, although I was already ill by this time. All of my efforts pertaining to aviation, rocketry, and inter-planetary travel, I bequeath to the Bolshevik Political Party and the Soviet government, as a genuine guidance in the progress of humanity's culture. I feel certain that this gesture will successfully conclude the work I have started.

A response came quickly:

> To the famous activist of science, comrade K.E. Tsiolkovsky:
>
> Accept my gratitude for the letter and the complete assurance of the Bolshevik Party and Soviet government. I wish you health and further productive work on behalf of all the workers in our country.
>
> I firmly shake your hand.
>
> J. Stalin.

[1] Referring to the October Revolution.

On September 19, 1935, at 10:34 PM, Konstantin Eduardovich Tsiolkovsky died of cardiac arrest. He had just turned 78 years of age. An obelisk was erected over his grave with a plaque inscribed, "Humanity will not remain on Earth forever."

In later years Werner von Braun, after the success of the Saturn-5 program, expressed his view of how greatly the modern space program had depended on Tsiolkovsky's initial efforts:

> The results of his pioneering efforts are obvious to all, to all those who today work in the sphere of cosmonautics. He left for us these mathematical equations that are indispensable for understanding the problems tied with the construction of multi-stage rockets, and especially pertaining to the design of the Saturn-5 rockets.
>
> The calculations Tsiolkovsky provided several decades ago have not to this day lost their significance. His theories have been proven over time.

Konstantin's wife Varvara Evgrafovna passed away five years later on August 23, 1940.

Konstantin Tsiolkovsky: Translation of Selected Compositions

Autobiographical Writings

Sketches of My Life (1935)

Introduction

By nature or by character, I am a passive revolutionary and socialist. My composition, *Sorrow and Genius*, published in 1916 when the tsar was still ruling, serves as proof. In it I fully define and formidably propagate the advantages of socialism in the wide sense of the word.

But why did I not evolve into an active revolutionary? The reasons are the following:

Deafness since the age of 10, which caused me to be physically weak and handicapped.

As a result of this, the absence of friends, associates and community ties.

This also caused a deprivation of many of life's joys and my financial helplessness.

To escape from the above dilemmas, I reformed my aspirations, and turned to technology, science, research and natural philosophy. Initially this was in the sphere of dreaming, and then my innovations surfaced. This was the reason why so many conceited and dedicated scholars rejected me. I was an upstart, reformer, and as a result I was not recognized.

It is not of my own will that my biography consists of life's trivialities and my efforts. Everybody consumed the latter, while the former, all that remains, is mundane and worthless. Other than this, due to the fact that I am limited by life's impacts, my biography cannot be as embellished as normal persons who have no physical handicaps.

Of course, there do exist several biographies of myself, published in magazines, in sections of books, or in the form of introductions to my compositions. They are not bad, but some are embellished or biased one way or another. You will only notice the mistakes in them if you compare them to my autobiography. So no matter how bad it is, is it still a useful source that will shed light on my life and activities, and from another point of view.

Genealogy

In the matter of the progress of humanity we seldom notice the influence of our heritage. All of these Faradays, Edisons, Fords, Grahams, Columbuses, Watts, Stevensons, Neutons, Laplaces, Franklins, and others, evolved from among their own people and did not have any talented ancestors. There are no traces of heritage here to be noticed. It is clear that genius is more created by conditions than what is transmitted from the parents or other ancestors. Maybe the ancestors did have talents, but they were not displayed worldwide. Their impression was meager.

Only in very rare circumstances are talents clearly transmitted due to inheritance. So the son of William Herschel and the son of Charles Darwin were famous, but distant from their fathers. Such examples in history are much smaller than the reverse.

Nevertheless, we must not reject the influence of inheritance in its entirety. In childhood and adolescence this did not interest me at all and so I never bothered to research it in my own family. But then my deafness also interfered. My mother had some Mongol ancestors and her maiden name was a Mongol derivation. At the same time I could not understand the value a person would place anyway on the significance of genealogy. I did hear a rumor that my father was related to the popular [Severyn] Nalyvaiko, and that my father's earlier forebears carried this name.

Based on family tradition, the progenitor of the Tsiolkovskys was the popular revolutionary leader Nalyvaiko. According to the encyclopedia, he was a Cossack leader at the conclusion of the XVI century, a fighter against the Polish aristocracy. He was born in the city Ostrog.

My father's character was near to being choleric. He was always cold, restrained, and never did he argue with my mother. My entire life I was only a witness to one occasion of an argument with my mother, and in this case she was the one to blame. He did not respond to her arrogance and did not want to divorce her. She asked for forgiveness. This occurred about 1866, when I was 9 years of age.

Among his associates he had a reputation of being an intelligent person and orator, while among officials he would blush and was intolerant due to his high moral standards. He smoked a lot, and his vision was poor his entire life. I remember him as being far-sighted. He would wear glasses when reading. As a young man my father occasionally drank alcoholic beverages, but once I started to grow, he put this aside. He seemed to have a gloomy view of matters and seldom did he laugh, and was terribly critical and always willing to debate. He never agreed with anybody on anything, but never lost his temper. He was distinguished from those surrounding him due to his strong and difficult character. He never picked a fight and never insulted, but everyone else seemed to pressure him. We were afraid of him, although he never permitted himself to irritate someone, or use profanity, much less fight.

Was my father a know-it-all? Relative to this era his education was no less than the surrounding society, although, as the son of poor parents, he barely knew the [Russian] language and only read Polish newspapers. While young he was an atheist, but in his later years he would attend a local church together with my sister. However, he was far from entertaining any clergy. I never saw a Polish Catholic priest or an Orthodox priest in our home.

He did have some passion for invention and building. Before I was born, he thought of a design for a thresher and built it, but it was a failure. My older brother related to me that he together with them built miniature houses and palaces. He encouraged us with every physical labor and in general to be self-sufficient. We did almost everything that needed to be done on our own.

My mother was of a completely different character: she had a sanguine nature, hot-headed, but also laughing and joking, and very talented. My father possessed a character of a strong will, while my mother was the talented one. I really liked her singing. My mother married at 16 years of age and my father was her senior by 10 years.[1] My parents loved each other very much, but this was not always obvious. My father also enjoyed the company of other women, but he was never unfaithful.

My parents had a predilection toward attire, toward superficial appearance and a respect for cleanliness and humility, especially my father. During winter we wore inexpensive coats, while summertime and in our home we only had our shirts. It seems that we did not have any other clothes. The exception was attending school, and then I had a nicer coat to wear.

[1] A minor typographical error on the part of Tsiolkovsky, the difference was 8 years.

Father never spent time in prison, but he did have altercations with the police and many unpleasant disagreements with government officials.

The government recruited my father to work in some forest related business, where he spent 5 years. He used his experience to then teach natural science at some local school, but this only lasted a year. Then he become some lower level official, and in general did not rise up the ladder, but his career seemed to go down. Eventually the provincial government assigned him back into the forestry division, but this only lasted a few months, as the minister did not confirm his assignment. Now we were back into destitution.

Father was always healthy; I do not remember him being ill. He ate moderately and never was fat. He was heavy set, but he did not have a stomach, and was of medium height and rather muscular. He did have a full head of hair and was a brunet, although gray was starting to show. Toward the end of his life he became melancholy and depressed, although never complaining, and did not leave the house to go anywhere. Then he suddenly died, and he was not even ill. As an aunt explained it to me: he rose in the morning, sat for a while, sighed a few times and that was it. I had just started my teaching career when this occurred. He was 61 years of age when he died.

Mother was always in good health. I never saw her lying in bed and never saw the blanket over her head, but she always seemed to be having labor pains and suffering from childbirth. She gave birth to 13 children.[1] My last brother died 20 years ago, and my last sister about 15 years ago. She had one daughter, my niece, who is still alive, and one of my brothers had children. Mother was above medium stature, brown-haired, with a few distinguishing Mongol features.

How can I describe those qualities of my parents that I inherited? I think that I acquired my father's strong will and my mother's domestic talents. Can I say this about my brothers and sisters? Yes, because they seemed to be normal and happy. I felt inferior due to my deafness, poverty, and lack of basic necessities those years. It later haunted me, but compelled me to work and search.

Birth—1857

From what I was able to discover, my parents' mood before I was born was cheerful. This occurred in 1857, before the liberation of the serfs, and which was increasing the excitement of the community.

[1] Of the 13 children, only 6 are actually known about, the balance no doubt died at birth, or else a typographical error on the part of Tsiolkovsky, since he wrote this at the age of 77, in January 1935, a few months before his death. Eduard and Maria were married 21 years.

September 4, 1957, was a nice day, although cold. Mother took my two older brothers, ages 5 and 6, for a walk. When she returned labor pains started and on the next day a new citizen of the universe appeared, Konstantin Tsiolkovsky.

First Impressions (ages one to 10, the years 1857–1866)

I imagined this as a dream, that a giant was leading me by the hand. We descend a staircase to a flower shop. With fear I stare at the giant. I think that this was probably my father.

From the ages 3 to 4: my mother receives letters. My grandfather—her father—dies. Mother sobs. I look at her and start to cry also, but they put me to bed. This was during the day. I look at animals in Daragan's book.[1] The picture of a walrus scares me and I hide under the table. I watch my father write, and it seems to be something easy to do and I give it a try.

Ages 5 to 6 years: I do not remember who showed me how to print the alphabet. For every letter I learn, my mother gives me a kopek. I enjoy riding the sleigh on skis, and every little effort causes the sleigh to move. It feels so wonderful. I cannot forget such a joyful feeling the first time I saw water in the pond. Toys were not expensive, but I had a habit of breaking them to discover what was inside.

Ages 7 to 8 years: Someone made a gift to me of a copy of Afanasyev's tales.[2] I started to study them, became interested and so taught myself how to somewhat read.

I had the measles, it was springtime. I felt wonderful when I recovered. My parents and guests at our home liked me when I was small. They called me by various nicknames: little bird, happy kid. Once I stole money from the table. I went to bed without tea, and cried for a long while, feeling sorry for myself.

We were not afraid of our mother, even though she would pull our hair, but it was not painful. But father impressed fear into us, although he did not spank the smaller siblings and did not cuss. Never did he lose his temper or shout at us.

Ages 7 to 9: Grandmother died and mother went to her village for the funeral. We remained alone. I was lonesome for her and even became depressed.

My older brother vexes me. I chase after him and throw rocks. Father gets involved, "What happened?" "He hit me in the head," brother Mitr

1 Anna Mikhailovna Daragan (1806–1877), author and illustrator of children's books.

2 Aleksandr Nikolaevich Afanasyev (1826–1871), Russian folklorist and ethnographer.

tells him. My father whips me twice with a tree branch and was it painful. I feared these lashes as I did fire, although I never received any more than 2 or 3 lashes. Father was a just and humane person, but I could not reconcile this with being lashed.

So this was the era. When my father attended a Jesuit school in Volyn, he was whipped—not every day, but at least twice a week. I was whipped only 5 times my entire life. Is not this progress? I did not retain any contempt for my father or mother as a result. Of course, I was not an advocate of corporal punishment, much less the rod, but we need to take into consideration the era, and even kings were lashed at this time.

We dug a well. Before water filled it, we—the children—climbed down into the well. It was such a curious place. The mountains surrounding us were so beautiful during winter. This was my first opportunity to go ice skating and riding on a sled.

During the summer we built sheds. It felt good to do some creative work. Sometimes we even built bathhouses. In autumn we would heat them and stay warm. It was my personal Camelot.

My education was difficult and painful, although I was capable. Mother helped us. Father also attempted some pedagogic assistance, but he was impatient and did more harm than good. I remember when someone brought an apple and poked some sticks into it. This was supposed to be our Earth with a rotation around its axis. The teacher became irritated, called us all stupid and left. Someone ate the apple.

I am assigned the task of writing a couple of lines on a slate with chalk. I become bored with the exercise. But when the torment ends, I feel such satisfaction from relief. Once, mother explained to me how to divide whole numbers. I could not understand it and just listened, and ignored it all. Mother became irritated and gave me a good slap. I cried, but then I understood. I came to the conclusion that children should not be beaten. We need to find better means to gain their attention.

I passionately loved reading and read everything that we had and whatever I could get my hands on.

I loved to dream and even paid my younger brother to listen to my fantasies. We were small and I wanted our house, people and animals to also be small. Then I dreamed of physical strength. I fantasized about jumping high, climbing high like a cat, climbing up ropes. I dreamed of a complete absence of gravity.

I loved to climb onto equipment, on roofs and in trees. I would jump from equipment to try to fly. I loved to run, play soccer, lapta,[1] gorodki,[2] blind

[1] A primitive type of baseball played in Russia.

[2] A Russian game similar to bowling except sticks are thrown to knock over pins.

man's bluff, and others. I would grab a snake by a stick and toss it into the air, and gather spiders into boxes.

We had a large meadow in our yard that would fill with water after the rains in autumn. The water and ice would lead me into a mystical mood. We built a trough and attempted to float in it, and in the winter we made ice skates from wires. I made them myself, but hurt myself on the ice so bad that I saw stars in my eyes. Finally I was able to get some used and damaged ice skates somewhere, and I repaired them. I learned to ice skate in one day. I was even able to ice skate through town on them to the apothecary to buy something.

This was the period of my normal childhood until I went deaf at 10 years of age. In no manner was it different from the life of any other ordinary child and I do want to underscore this. The conclusion is interesting, but not anything new. A person can never guess what the future may bring.

We love to embellish the childhood of great people, but this is not entirely artificial, but the invention of a preconceived opinion.

Sometimes what occurs is that future famous people display their capability very early and their contemporaries just guess at their great destiny. But in the immense majority of incidents, this does not occur. Such is the truth confirmed by historical examples. Nonetheless, I personally think that the future child cannot be determined beforehand. Some talents develop and surface in childhood, yet later in life they do not provide results.

Deafness (from ages 9 to 11, years of 1866–1868)

From this point forward is the biography of a person not considered normal, but semi-deaf. It will not be very brilliant, as distinguishing events are meager as a result.

The beginning of winter during age 10, I was riding a sled in the snow, but then caught a chill and it developed into scarlet fever. I became quite ill and had convulsions. They thought I would die, but I recovered, except that I lost most of my hearing and I did not recover from this. It tortured me. I would dig in my ears with my finger thinking it would help by releasing something or other, but I think I only made it worse, since once I caused bleeding from my ears.

The results of the illness: the absence of clear sounds, sensations, communication with people, and a personal humiliation, caused me to become callous or insensitive to my surroundings. My brothers attended school, but it did me no good.

From what I later discovered, the deaf can still effectively learn, but from personal study, since they cannot hear the teachers. My father told me about himself, that he did not intellectually develop until he was 15 years of age. Maybe this was partly a reason for my own later development.

Period of Indifference (ages 11 to 14, the years 1868–1871)

Being deaf causes my biography of the near future to be rather uninteresting, since it deprived me of association with people, auditory observations and deductions. There was a dearth of persons involved in my life, but at least I did not get into any altercations. It caused me to be introverted. This is the biography of a handicap. I will relate to you conversations and describe my meager associations with people, but they will not be reliable or complete. In time, my hearing improved a little, and I remember more of what occurred after this.

I will bring to your attention one character trait and perhaps weakness. While in Ryazan, I met a boy older and stronger than me in the street. This boy thought of himself as a rooster. Immediately we took up a posture ready to fight. At about this time my cousin was walking by; he was rather robust and he intervened. "So what should I do to him, Kostya?"[1] he asked. I answered, "Don't touch him." This boy vanished. Then I realized that I was somewhat cowardly and never felt any need for vengeance. I was afraid of boys attacking me and especially criminals. I was also afraid of the dark.

I had an aunt who would tell stories of what would happen on dark nights and this would scare me. My father considered all of this nonsense, and so my aunt would not tell us this stuff in the presence of my father. What really scared us were the stories about cholera, war and other disasters. Of course, this was because we were children with undeveloped feelings; courage grows with maturity.

I had an inclination toward lunacy. On occasion at night I would rise and would babble unconsciously for a long while. Occasionally I would leave my bed and wander through rooms and hide behind a sofa. Once my parents came into my room and did not find me in my bed, and I was sleeping on the floor in another room. My brother Mitr had this malady even worse.

We would play dominoes and cards. I am glad for this, but now I have acquired a repulsion for playing cards and gambling, checkers, and even chess and similar table games.

Thanks to a good friend, my father was assigned to some kind of small job in the forestry division in the city Vyatka. Nearby was a beautify river that was so full of water. We swam there during summer and here I learned to swim. We took advantage of the freedom, walking where we wanted. I am now surprised that I did not drown in this river. Once, this almost happened,

[1] Diminutive of Konstantin.

although not when we were swimming. It was flooded with snowmelt at the time and ice would float and then stop. The day was beautiful, sunny. I want to ride on the ice floes. They would float near shore and it was easy to jump on one. So I and a friend descended the slope to the shore, and we jumped from one ice floe to another. Between a couple of the ice floes was some dirty water that I mistook for a dirty ice floe, and I stepped there and fell in. Due to the cold I opened my mouth. Then my friend ran to me to help and jumped into the ice bath and pulled me up and onto another ice floe. Somehow we made it to shore all wet and ran home to dry.

There was a large park in the city. In it was an immense swing that would hold 10 persons. It was a very heavy box on ropes with benches. I decided to ride on this swing. After swinging for a while, I could not hold to it any more and I fell into the grass and almost broke my back. For a while I laid in the grass, rolling around from the pain. I am thinking that I am dying. After a short while I felt better and went home with my brother. I had no further consequences. Eventually, the swing was removed.

A year before mother died, my parents, and especially my mother, were struck unexpectedly by the death of my brother at age 17. Two of my older brothers were studying at the time in Petersburg, and the younger of them died of delirium tremens. The sorrow of my mother is indescribable, and this was sadder for us than the death of our brother.

When I was 13 years of age, we lost our mother, and she was not yet 30 years of age.[1] This is what happened. After we had tea in the morning, mother said to me and my younger brother (he died very young), "Will you weep if I die?" We responded with bitter tears. After this mother became ill, and then critically, for a while, and then died. Before she passed away she called us to herself and bid us farewell. Mother then lay unconscious and we watched tears coming from her eyes. I wiped them off her face and wept.

But sorrow does not seem to leave an impact on adolescents, not profound and devastating. After a week I was back to climbing a cherry tree, and swinging with pleasure on a swing.

Mother's younger sister took charge of our household after her death, but we did not especially like or respect her. But she was nonetheless meek and never treated us badly: not one sound [of reproach] and not one shove. She did have an inclination to exaggerate things and even lie, and she had absolutely no toleration for idleness, which we did not like.

In our city there was an ancient but rather tall church. At the top of it was a tower with a balcony, and it reminded me of a lighthouse. Perhaps in earlier eras it had served as a lighthouse. During Easter Sunday boys would crawl into the bell tower to ring the bell. I did the same, except not

[1] Another typographical error; Maria was about 38.

to ring the bell, but just to climb into the balcony. The view from there was beautiful. I was alone. No one dared to climb up there. I considered this an immense satisfaction: everything was below my feet. I would sit for a while, I would stand for a while, I would walk around for a while. Then I thought about shaking one of bricks in the wall. But not only did it shake, but the entire balcony seemed to shake. I was frightened, imagining myself falling from such a terrifying height. It seems that my entire life passed right in front of me as I rocked in that tower. Then I regretted having climbed up so high.

We did not have a governess, nurse, or maid, and of course, we could not. Close associates would pity me for my condition, but there was nothing they could do. Mother died, father was consumed with making a living, our aunt was little educated and weak.

This 3-year interval, due to my indifference, was the most melancholy, darkest period of my life. I am striving to bring some of it to memory, but right now I cannot seem to recall anything. Looks like there is just nothing worth recollecting. All I remember is riding in the streets on sleds, ice skates and sleighs.

Flashes of Awareness (ages 14 to 16, the years 1871–1873)

During the 11 years we lived in Ryazan, I enjoyed building figurines, houses, sleds, pendulum clocks and other toys. All of these were made of paper and cardboard and glued together using wax. My inclination toward handicrafts and articles surfaced early. With my older brothers it was more intensive.

During the ages of 14 to 16, my need for manufacturing appeared in a higher manner. I assembled self-operating carriages and locomotives that moved from the force of a spring that was tightly wound. The steel I acquired from the hoops in hoop dresses that I bought at local fairs. This really impressed my aunt and my brothers even took notice of what I was doing. At the same time I built tables and boxes that we used to store stuff.

Once I saw a turning lathe, and decided to build my own. I built it and carved a figure from a piece of wood on it. My father's friends said that nothing would come of it. I carved a lot of windmills of different types on it. Then it was a wagon with a windmill that would move against the wind and in any direction. This even impressed my father and he started to think highly of me. After this I built a musical instrument with one string, one keypad and a short bow that would move quickly across the string. Then I connected it to a pedal that was moved by wheels. My next project was a large windmill on a sled for riding. I started on it but then realized that we did not have enough or constant force from the wind to move it, and I abandoned the project.

All of these were toys that I manufactured all by myself, independent from any study of science or technology books.

It was during my studies that flashes of a serious intellectual cognizance appeared. At the age of 14 years, I suddenly decided to read a book on arithmetic, and it seemed to me that everything there was completely clear and understandable. From this time I realized that books are things not only for the scholar and they were fully accessible to me. With curiosity and interest I devoured several of my father's books on natural sciences and mathematics, those which he had collected when he was teaching classes on surveying when he was working in the forestry division. What attracted me was the astrolabe, the measurement of distances to inaccessible objects, and drawings charts and defining altitudes. I designed an altimeter. With the help of the astrolabe, and not leaving the house, I determined the distance to a lighthouse. I estimated it to be 400 yards. Then I went to ascertain it. Guess what? I was right. So did I confirm theoretical knowledge.

The study of physics motivated me toward the manufacture of other equipment: an automobile driven by a jet of steam, and a dirigible filled with hydrogen that, understandably, failed. Later I manufactured a sample machine with wings.

I remember one incident toward the end of this period. My father had an inventor-friend, an educated forestry official. He came up with the idea of an eternal motor, but he never fully accounted for the laws of hydrodynamics. I spoke with him about it and I immediately understood his error, although I was unable to convince him. My father believed him. Then in Petersburg they wrote of his successful invention in the newspapers. My father advised me to be modest and accept it, but I stood by my opinion. I was glad I did this as it impressed upon me for the future to stand for principles.

Essentially, nothing extraordinary during this period occurred in my adolescence. What I have written is as much as happened, although not very brilliant.

In Moscow (ages 16 to 19; the years 1873–76)

My father speculated that I had some technical abilities and so sent me to Moscow. But what was I supposed to do there, with my deafness? How could I connect with anybody or anything? Without a knowledge of life I was also blind in regards to a career and work. I received 10 to 15 rubles a month from home. I lived on black bread, and could not afford potatoes and tea. But I did buy books, mercury, chemical glassware, sulfuric acid and such items.

I distinctly remember that, other than water and black bread, I had nothing to eat. Every 3 days I went to the bakery and bought 9 kopecks of bread. In this many, I lived on 90 kopecks a month.

My aunt knitted me a bunch of socks and sent them to me in Moscow. I decided that I would do fine without them, and what a mistake this was. I sold them cheaply and then used the money to buy alcohol, zinc, mercury, sulfuric acid and more of such items. As a result of the acid, I walked about in pants with yellow spots and holes. Boys on the street would notice me and say, "So did mice eat your pants?" I also walked about with long hair plainly because I did not have the time for a haircut. I was horribly confused in basic areas, but I was nevertheless happy with my ideas, and the black bread did not cause my life to be bitter. In fact, it never dawned on me that I was hungry and losing weight. But exactly what was I doing in Moscow? Was it just limited to pathetic physical and chemical experiments?

The first year I diligently passed a systematic course of elementary mathematics and physics. Often, reading some theorem, I sought my own independent proof, and I liked doing this more and it was actually easier than following the explanation in the book. But I was not always successful. This was obvious as a result of my inclination to independent evaluation.

During my second year I studied higher mathematics. I passed my self-instruction in higher algebra, differential equations, integral calculus, analytical geometry, spherical trigonometry, and others. Various questions seemed to occupy my thinking formidably, and I aspired to immediately apply the knowledge I acquired to the resolution of these questions. So I almost independently studied analytical mechanics.

Here for example are some of the questions that occupied my time. Of course, many questions that surfaced were resolved before my familiarity of higher mathematics, and so some of them were later decided in the opposite.

> It is possible to practically utilize the energy from the rotation of Earth? The decision was correct: negative.
>
> What form does a liquid maintain on the surface of some vessel when it is rotating around some vertical axis? The response was accurate: the surface is a parabolic stream.
>
> And since telescopic mirrors have this form, then I thought to build a gigantic telescope with moveable mirrors, out of mercury.
>
> Is it possible to build a railroad train encircling the equator, so that due to the centrifugal force, gravity would have no effect? The answer is negative: the resistance of air interferes, and other items.

> It is possible to build a metallic dirigible, not releasing gas and eternally suspended in the atmosphere? Answer: it is possible.
>
> It is possible to utilize released steam in steam engines requiring a high pressure? My answer: it is possible.
>
> Is it possible to apply centrifugal force to propel something beyond the atmosphere, into outer space?

And I thought of such a machine. It consisted of an enclosed capsule or box, where 3 solid but elastic vibrating pendulums were attached upside down, with spheres connected at the upper vibrating ends. They are designed as arcs and the centrifugal force of the spheres had to raise the cabin and carry it into outer space.

I was in such ecstasy that I could not sit in one spot and went to spread my inherent joy into the street. I wandered about Moscow an hour or two in the night, deliberating and pondering my discovery. But alas, while walking along one street I realized that I was mistaken: all that would happen is that the machine would shake and nothing more. Its weight would not decrease by even one gram. However the temporary ecstasy was so strong that for the balance of my life I saw this mechanism in dreams and I would ascend in it with great enthusiasm.

But did I really have absolutely no friends in Moscow? I had some accidental acquaintances. So there was a student B., whom I met at the Chertkovski Public Library, and what interested me about him was that he was completing the course in mathematics. He was at my residence once or twice and wanted to read Shakespeare to me, which I enjoyed very much. But when I became old, I decided to reread him, but then I discarded the attempt as being an arbitrary effort. (And Leo Tolstoy said the same about himself.)

Another chance associate wanted to introduce me to a girl. Up to this time my stomach was filled with solely black bread, and my head with wild and fantastic dreams. But even under such conditions, I did not escape supra-platonic love. So this is what happened. The maid at my residence also washed clothes at the house of a popular millionaire, Tz. While she was there, she talked to them about me, and the daughter of Tz. became interested in me. The result was a long series of letters between us. Finally she stopped writing due to circumstances beyond her control. Her parents found her letters and felt them to be suspicious, and then I received a final letter terminating the relationship, although I never did discover the author of this final letter. This did annoy me for a short while, yet it did not interfere with me falling in love again in the future.

It's interesting that in one of my letters to her I affirmed my position of being such a great person, such a one that has never been in the past, and there will not be one like me in the future. In a return letter the girl laughed at this, and now it bothers me when I remember her words. But what self-confidence, what boldness, I possessed at the time, especially in view of my pitiful situation. True, I did think about a triumph over the entire universe. Occasionally I would remember the cliché, "Pity the soldier who maintains no hope of being a general." However, such fantasies passed in my life and without leaving a trace.

Now on the contrary another thought torments me. Did I ever pay with my efforts for the bread I ate over the course of 77 years? This is why my entire life I aspired toward a peasant's manner of agriculture, in order to literally eat my own bread. My ignorance of life interfered with its materialization.

What did I read in Moscow, and what did I do in my spare time? First of all, I studied sciences. What I avoided was anything vague or philosophic.

The popular young journalist Pisarev[1] compelled me to shake with joy and happiness, because in him I saw a second me. In later years I looked at him differently and noticed his mistakes. Nevertheless he was one of my most respected mentors. I was also attracted to other editions of Pavlenkov.[2]

As far as best sellers were concerned, Turgenev made the greatest impression on me, and especially *Fathers and Sons*. However, in my later years, I reevaluated his books and did not rate them so highly. I also read much of Arago[3] at the Chertkovsky Library along with other books on science.

By the way, at Chertkovsky Library I noticed a worker with an unusually happy face. Never up to that time had I met anybody like him. Obviously it is true that the face is the mirror of the soul. When exhausted and homeless people would migrate into the library, he did not seem to pay attention to any of it. This would have annoyed any other librarian formidably. He would offer me forbidden books to check out. Then I realized that this was the well-known ascetic Feodorov,[4] a friend of Tolstoy's, an amazing philosopher and a most modest person. He distributed all of his meager income to the destitute. Now I look back and see that he also wanted to contribute something [monetarily] to me, but he did not succeed, as I shunned his attempts.

Then I also found out that for a while he had been a school teacher in Borovsk, where I would later teach. Due to his humility, he did not want any of his books to be printed, regardless of their promising popularity and the urging of friends.

---

1 Dmitri Ivanovich Pisarev, 1840–1868.
2 Florenti Feodorovich Pavlenkov, 1839–1900. A Russian book publisher.
3 Francois Arago (1786-1853. A French mathematician, astronomer and physicist.
4 Nikolai Feodorovich Feodorov (1829-1903. Russian librarian, and futurist and cosmic philosopher.

Again at the City Vyatka (ages 19 to 21; the years 1876–78)

I periodically corresponded with my father. I was happy with my dreams and never complained. But father felt that such a life in Moscow would exhaust me and bring me to ruin. The remaining family invited me to return to Vyatka with plausible propositions. They were all happy with my return, but were surprised at my sullenness. In plain words, I lost much weight.

In the liberal sector of society my father took advantage of the respect and had many associates. Grateful for this I was able to give lessons. I had success and soon I was flooded for requests for lessons. The high school students advertised my abilities, as though I explain algebra in an understandable manner. But never did I ask for money or count the time that I spent with them. I took what they gave me—from a quarter to a ruble per hour. I remember one physic's lesson, and they paid generously—a ruble an hour, and the student was very capable. When in geometry we reached the topic of complex polyhedrons, I did a good job in gluing together cardboard to make the shapes and used them in my mentoring.

When in physics we reached the topic of aerostatics, then I glued together a lot of cigarette paper making a dirigible about a yard long and went with it to my student. The flying montgolfier amazed the boy.

It was only in Vyatka that I accidentally discovered that I was near-sighted. I was sitting with my younger brother on a river bank and we were watching a steamship. I was not able to read the steamship's name, but my brother, wearing glasses, was able. I took his glasses and also read the name. From this point on, I wore glasses with concave lenses, and I still wear them; yet I always read and still read without glasses, although right now I seldom read. Seldom do I have to condescend to using a magnifying glass.

In Moscow I would walk about in my elder brother's coat, which was a remodeled version of my aunt's tattered overcoat. It was too large for me, and so to hide this I would wear it slung over my shoulders, regardless of the occasional infernal cold. The coat was made of some basic thick woolen cloth, although without a collar and lining. But it was soon stolen from me when I was walking once near the Apraksin market. Some boys jumped out and almost forced me into the market. They took my coat and tossed me one that was in tatters, and then I gave them 10 rubles to leave me alone.

Another unsuccessful event was when I bought some shoes at Sukhoverk. I lost my old ones and returned home with some shoes that had no soles.

At the city Vyatka, I was hired to design some equipment at a local machine and equipment manufacturer. I was also able to rent an apartment that had an independent workshop.

Among other things, I built something like water skis, with a high scaffold, a complex arrangement of paddles and a centrifugal driver. I was able to successfully traverse a river. I was thinking I could gain some speed, but made a crude mistake: the skis had a weak stern causing much resistance, and so I was not able to gain much speed.

My brother, a year younger than myself, caught a cold and became ill. He was the one I was closest to from early childhood. The winters here were freezing. My brother lost his appetite; he developed ulcers in his intestines and then died. His friends at the high school he attended participated in his funeral. I declined to attend, feeling that there was nothing I could do for someone who had died. This action of mine was not the result of the freezing weather, but because of my intense sorrow. Then I realized that people attend funerals for the sake of relatives and friends.

I lugged around books and magazines from the public library, those on science. I remember the inventions of Weisbach[1] and Brashman,[2] and Newton's *Principles*, and others. During these years I reread magazines dealing with Russian literature and national affairs several times. The influence of these magazines on me was immense. For example, reading the articles against tobacco, I did not smoke my entire life. I also developed a repulsion for European cooking and food. I was in pain my entire life, but I do not remember taking any efforts toward getting well. I only comprehended the great future of medicine when it was already too late. Hygienic articles had quite an influence on me. A repulsion for the alphabets of other countries surfaced during my studies.

Settlement in Ryazan (ages 21 and 22; the years 1878–79)

Father became ill. The death of his wife and children, and life's failures, contributed to it. Father retired from work with a small pension, and so all of us decided to relocate to Ryazan, to where we started. We took a steamship there during the spring.

In Ryazan we moved into the same house where we used to live, but now everything seemed to be very small, pitiful and dirty. Our previous associates and the local residents were now old. The gardens, yards and houses no longer seemed as interesting as before. I developed a sudden disappointment with the old places. In 1878, I was not yet a schoolteacher, and finally I was conscripted into the recently-legislated obligatory military service, postponed now for some time. I had a repulsion and contempt toward the concept of war, but at the same time I understood that it was difficult to go against the grain. Of course, not one person expected me to successfully complete my military obligation. As a result of my deafness, what occurred was an inevitable series of comic scenes.

1 Julius Weisback (1806-1871), a German mathematician and engineer.
2 Nikolai Brashman (1796-1866), Russian-Austrian mathematician.

I stripped half-naked while someone held my shirt. I told them of my deafness. The doctor listened as air passed through my ears as I was breathing. I do not remember exactly if they immediately discharged me or postponed it for a year. I only remember that the governor remained unsatisfied with the decision of the enlistment committee and wanted all those who had been released to be again examined.

He asked me, "What do you do for a living?"

My answer, "I teach mathematics." This caused him to ironically shrug his shoulders, and which just further confirmed my worthlessness as a soldier.

I remember that about this time I made an experiment using chicks. On a rotating centrifugal device I was able to increase their weight by 5 times and this caused absolutely no harm to them. I had done the same kind of experiments earlier in Vyatka but there I used some insects. I likewise subjected myself to experimentation: for several days I did not eat or drink. I was only able to deprive myself of water for the course of two days. What happened a few days later was that I lost my eyesight for a few minutes, and then I stopped the experiment.

The following year I passed the examination to become a schoolteacher, since here in Ryazan I was unable to do enough tutoring and so lived on a very meager income. I moved from our house and into a room I rented from a worker named Palkin. He was Polish and had earlier been sentenced to Siberia for some reason, but now he was released.

I was afraid of being tardy for the examination. I asked the sentry, "Has the examination started?" He laughed at me and answered, "They are only waiting for you."

The first verbal examination was on Catechism. I was totally lost and was not even able to say one word. They took me aside and sat me on a sofa. After 5 minutes I recovered my composure and replied without stumbling. After this scenario, I did not lose any of my faculties. What was primary was that my deafness interfered with my ability to answer the questions. I would answer absentmindedly and had to often ask what the question was, and this bothered me. The written examination was in the director's office and in his personal presence. After a few minutes I wrote a composition providing completely new proofs to the science questions. When I gave it to the director, his question was, "Is this a draft?" I answered, "No, this is the final."

It was good for me that I was assigned a reasonable and young examiner. He gave me a better grade than I deserved, and did not make even one comment on any of my answers. I do not know my final grade, only that a grade less than 4 is failure.

During my verbal examination, one of the teachers consistently picked his nose. Another one, examining me on Russian literature, was writing something or other the entire time and he seemed unconcerned about any of my answers.

My father was very content with the results and he decided to help me with supplies for my assigned teaching position. I took the examination wearing a gray patched shirt, while my coat and pants were in a pathetic condition, and I had almost no money remaining. I had a tailor sew me a new uniform with pants and coat for only 25 rubles. In fact, for the next 40 years of my teaching career, I never had another uniform. I never wore a cockade and I wore whatever I felt was comfortable for me. Neither did I have a starched collar. I was able to get an inexpensive coat for 7 rubles. I had earmuffs sewn to my hat and I was ready for the cold weather. Then I insulted my father when I returned to him my worn-out and used clothes.

Regardless of my applications, I was assigned a teaching position that would not start for another 4 months. I spent this interval of waiting in a local village at the home of a landlord, Mr. M. I was a tutor for his children. I taught them grammar. The boy would ask me, "Why is a hard sign placed at the end of words?" I answered, "Due to stupidity."[1] Likewise I did a critical evaluation of the entire grammar of the Russian language. When the boy would notice a hard sign he withdrew into a state of torpidity, but then acknowledged, "I now know this is stupid."

Grammar was an amusement for me. What was most important is that I submerged myself into the laws of gravity of various masses and studied the various types of movement, and which were related to gravity. It was 30 years later that I sent the remnants of these notes and sketches to Perelman,[2] as historical documents. Just recently, in 1932, he mentioned me in one of his books.

Every day I strolled quite a distance from my home and meditated on my experiments and on my concepts of a dirigible. I was warned that there were wolves in this region, someone pointed out their tracks to me, and even some feathers from killed birds. But the thought of danger never entered my mind and I continued taking my strolls.

At the Borovsk School (ages 23 to 35, the years 1880–92)

Finally after Christmas of 1880, I received news of my assignment as a teacher of arithmetic and geometry at the Borovsk county school. I put on my hat with earmuffs, overcoat, jacket, mittens, and departed on my journey.

[1] The hard sign ъ was removed from words ending in a consonant with the revision of the Russian alphabet in 1918.

[2] Yakov Isidorovich Perelman (1882–1942), Russian science writer.

Reaching the city Borovsk, I took a hotel room and then started to look for an apartment. The city was filled with Orthodox Old-Believer dissenters as well as many taverns. I was not one of the former and I never entered any of the latter. Houses stood empty, yet I could not rent an apartment. In one place I was able to find an empty and immense mezzanine, and I rented just one room in it. The first night I felt as though I was poisoned by fumes.

Later the room was rented to a newly-wedded couple who were friends of the owner and I was moved to a dark cell that I did not like at all. I started to look for another apartment. At the recommendation of local residents, I had lunch at the home of a widower and his daughter who lived at the edge of town, near the river. They rented me two rooms and a table, with two meals a day included. I was content here and lived here quite a while. The owner was a wonderful person, but he drank a lot. I was struck with amazement at his understanding of the Gospels. I conversed with him a lot and spent considerable amount of time during tea, lunch or dinner with his daughter.

It was time for me to marry and I married his daughter without being in love. I was hoping that such a wife would not interfere with my life's calling, but work on my behalf. This hope was fully realized.

We walked about 3 miles to the church to marry, on foot, not especially dressed for the occasion, and we did not allow anyone else into the church. We returned and no one knew anything about us getting married.

I never knew any woman in my life, before marriage or after, other than my wife.

I only attributed a practical significance to a wedding. A while back, I guess about 10 years before, I discarded all the stupidities of organized religion.

I was carried away by natural philosophy. I proved to my friends that Christ was only a good and intelligent person; otherwise he would not have stated things like, "He who believes in me will do what I do and even greater." What is important is not his exorcisms, healings and miracles, but his philosophy.

This was reported to the director in Kaluga. The director summoned me for an inquiry. I borrowed some money and went, but the director was at his summer house, so I went there. A kindly older gentleman met me and asked me to wait awhile until the director was ready to see me. I told him, "The taxi driver does not want to wait." This annoyed the director and when he appeared we had the following conversation:

"You want me to meet with you, but you provide me no money for travel."

"What happens to your income? How do you spend it?"

"For the most part I spend it on chemical and laboratory equipment, I buy books, I do experiments."

"You do not need to do any of that. Is it true, though, that in the presence of witnesses you said this and that about Christ?"

"True, but all of this is noted in the Gospel of John."

"Nonsense. This text is not to be found there and cannot be. What do you have in the way of money?"

"I have none."

"So how can you, destitute, be so bold as to make such statements?"

I had to promise not to repeat my errors and it was only by doing this that I kept my teaching position and had a job. Due to my lack of understanding of life, I had no other exit. This lack prevailed over me the entirety of my life and forced me to do what I did not want to do and so I tolerated much and was humiliated.

Anyway, I returned whole to my physic's experiments and my serious mathematical work. Electrical lightning flashed at my home, thunders echoed, bells rang, paper dolls danced, holes from lightning strikes occurred, fires flared, wheels turned, illuminations blared, and formulas brightly appeared. Simultaneously a mass of thunders struck me.

Among other items, I proposed to some people to taste a sample, a spoonful, of some invisible fruit jam. Deceived by my proposition, they received an electric shock. They enjoyed and were amazed at the electrical octopus that grabbed everything by their nose or fingers with its tentacles. Their hair would stand on end and sparks flew from every part of their body. Cats and insects likewise fled from my experiments.

I filled a rubber balloon with hydrogen and carefully balanced it with a paper boat filled with sand. Like a real creature, it floated from room to room, following the breeze, rising and falling.

At the school in 1882, my comrades called me a Zhelyabkoi,[1] and suspected something that did not exist. But I protected myself by participating in church services on every holy day and I took communion about every 4 years or so.

At this time I developed a completely independent theory of gases. Previously I independently studied a university course on physics published by Petrushevsky,[2] but there was only some indirect study on the kinetic theory of gases, and even then it was depicted as a questionable hypothesis.

I sent my treatise to the Physics-Chemistry Society in Petersburg. With a unanimous decision, I was selected as a member. But I never thanked them for this or responded at all, and due to my naive barbarity and inexperience.

[1] After Andrei Ivanovich Zhelyabov, a Russian revolutionary and an organizer of the assassination of Tsar Alexander II.

[2] Feodor Fomich Petrushevsky (1828-1904), Russian physicist.

I beat my head into the ground over the initial source of the sun's energy and I independently came to the same conclusions as did Helmholtz.[1] At the time there was no mention at all, or even a hint, of the radio-activity of elements. Then later compositions on this topic were published in various journals.

The river was near, but it was dreadful to float on a flat-bottomed boat, and we did not have any new boats. Then I thought of a special design for a high-speed boat. I rode on it with my wife who would sit at the rudder and guide it. A woodworker friend even won a bet with a wealthy salesman who said that I am incapable of building a boat. But then when I floated by his window in it, then he had to pay his loss. Then I even built such a boat to hold 15 persons. What happened next was that many others imitated me.

With the help of my boat, I threw a basket in the river and so caught fish. I got carried away with this and in early spring I caught typhoid fever as a result.

My boat's design was a surface of revolution that was shaped as a sinusoidal curve along its longitudinal section. Often we would ride on it with sails and never did we overturn the boat. Nevertheless the first model was very shaky and small and it scared people who would ride in it. They called it suicidal and dangerous.

During winter I would ski on the surface of the river with friends. This is what happened once. The water just froze and the ice was thin. There were 3 of us on skis and I was first in line. I told my friends, "If I fall through the ice first, then the two of you can ski back." The ice started to crack beneath me and water was showing through. I quickly fell to the side and crawled backwards and so saved myself. Was this boldness or stupidity? I think it was some of both. My friends ran to the village for help, but I was able to drag myself on my own to a more solid area.

How many times holding an umbrella did I speed across the ice during a storm due to the force of the wind! This was overwhelming!

I was always conniving or designing something. The river was near. I decided to build a sled with wheels. We all sat in it and used levers to turn the wheels and the sled was supposed to slide along the ice. Once it was put into motion the mechanisms just did not operate right and it was a failure. I doubted in the adequacy of its construction.

Then I exchanged this invention for a special sofa with sails and we slid across the frozen river. But the fluttering sails would scare the horses and the riders would curse at us, but because of my deafness, I could not guess as to what they were saying for a long while. Then after seeing the reaction of the horses, later I dropped my sails ahead of their appearance.

[1] Hermann von Helmholtz (1821-1894), German physician and physicist.

I would skate as long as the ice was thick and solid. Sometimes I fell into an ice-hole and get sopping wet, and the air was bitter cold. The water would flow from my coat and even create icicles on it. I would walk along the street and the icicles would hit each other making a ringing sound like bells.

I loved the river. Every day during good weather I would go for a ride on the river with my wife. She would guide the rudder and I would work the oars. When children came along, I would go by myself or occasionally with one of my friends. During autumn the seaweed along the bottom would wash away, and the water would be crystal clear. All the rocks, plants and animals living in the river would be visible. At times, I would swim along the current and watch all of this with great interest.

Blackberries grew along the shore in inaccessible places. The local region was beautiful, during the summer the river was damned and we would ride in our boat upstream and it was wonderful.

To be a schoolteacher was far from idyllic. The pay was small, the city was tight on their budget, and the students were not honest about completing their lessons. The teacher was often reprimanded by parents if the student did not understand the lesson.

I never catered to anyone, I was never idle, we never traveled and we were able to live on my income. We dressed plainly, essentially poorly, but never walked in tattered clothes and we were never hungry.

With my friends it was different. For the most part they were seminarians, finishing the courses and then taking the examination to be allowed to teach. They did not want to become priests. They were used to a better life, entertaining guests, taking vacations, squandering time, and drinking. Their income was not enough for all this, so they would take bribes, sold diplomas to the village students. I did not know any of this was occurring due to my deafness, and never participated in any of their orgies. To the extent I could, I did try to curb such conduct.

Regardless of my deafness, I enjoyed teaching. Most of the time I would share with the students the way to arrive at the answers. This motivated their brains betters and it was not so boring for the students to do this independently.

During the summer I would take the students of the older class to ride on my large boat, we would swim and practice geometry.

With my own hands I built two metal astrolabes and other instruments. I would take them when we traveled. I showed them how to draw plans, determine volumes and size of inaccessible objects and places. Nevertheless, most of the time was spent in frivolity and pranks, rather than actual work.

We had small family scenes and arguments, but I always admitted myself to be the one to blame and would ask for forgiveness. So was peace restored. We handled all the responsibilities: I wrote, did my calculations, soldered and brazed, drew plans, smelted, and more. I built air-piston pumps that worked well, steam engines, and various other equipment. Guests would arrive and ask me to show them my steam engine. I agreed, but only if the guest would hew wood to boil water.

For summertime I found another amusement for my students. I built an immense sphere from paper. At the bottom of the montgolfier was a screen made of thin wires, and since I did not have any alcohol, I placed several hot coals or burning splinters there. The montgolfier, its shape adjusting as it filled with hot gas, rose, but only as high as the string tied to it allowed it. But one time the string unexpectedly burned from the bottom and my balloon floated into the city, distributing sparks and burning splinters. It landed on the shoemaker's roof, and the shoemaker confiscated it. He wanted to bring me to court for this. Then the schoolmaster where I taught explained that I had released the balloon which fell on the house and then burst from the internal heat. So did they make an elephant from a fly.[1]

Then I took a different approach and only heated the montgolfier and extinguished any fire. It flew without a fire and then quickly descended. Children chased after it and caught it and returned it, so I would again release it into the air.

While I was 32–33 years old, I was concentrating on experiments dealing with air resistance. I then did some calculations and discovered that Newton's law on air pressure on an inclined plane was inaccurate. I came to other conclusions, less known at the time. I remember that during Christmas holidays I sat uninterrupted at this work the entirety of two weeks. Finally, my head was aching terribly and I quickly left and went ice skating.

I still have the entire manuscript that I wrote at the time. Later a part of it was published in a journal with the help of professor A.G. Stoletov.[2]

By the way, I need to say that I still have a copy of the original textbook on analytical geometry by Briot[3] and Bouquet[4] that I purchased in Moscow when I was living there. It seems that I still have other books of this era in good condition.

From the moment of my arrival in Borovsk, I was occupied zealously with the theory of lighter-than-air dirigibles. I would work on it during my

[1] The Russian equivalent of making a mountain out of a molehill.
[2] Alexandr Gregoryevich Stoletov (1839-1896), Russian physicist and professor at Moscow University.
[3] Charles Auguste Briot (1817-1882), French mathematician.
[4] Jean Claude Bouquet (1819-1885), French mathematician.

vacations, and I take no holidays to speak of. So it is now, as long as I am healthy and have my strength, I work.

It was also in 1887 that I met Golubitzky.[1] At the time the well-known Kovalevskaya was staying at his estate; she was a female professor from Switzerland.[2] He arrived in Borovsk in order to drive me to visit Kovalevskaya who wanted to also meet me. My poverty and handicap forced me to decide not to go, and maybe this was for the best.

Golubitzky then proposed that I go to Moscow to visit Stoletov [mentioned above] and read my report on dirigibles to the society. I went, and after wandering about the city, I finally found the professor. From there I went to the Poly-Technical Museum to meet with others, but I never did have the opportunity to read my manuscript. I only explained a few essential concepts and no one there objected. These people wanted me to move to Moscow, but nothing came of it.

Living at the edge of the city Borovsk, my residence was flooded. The waters rose half-way up the walls, all of our kitchenware was floating on top. We built a bridge out of chairs and beds and were able to float about. Ice floes struck the sides of the house. Boats came along to our windows, but we decided to just stay where we were.

We suffered seriously due to a fire on another occasion. Everything I had was either ruined or burnt. The fire started from a pile of hot coals in the neighbor's residence that had not been extinguished.

Once I came home from a friend's house. This was on the eve of a solar eclipse in 1887. There was a well in the street and something was shining near it. I walked near it and saw for the first time some brilliantly shining tinder or touchwood. I loaded a large armful and went home. I chopped them into small pieces and scattered them about the room. In the dark the impression was a star-filled sky. I invited whomever I could and they all enjoyed the spectacle. The solar eclipse was supposed to occur in the morning. It did occur but it was raining. I tried to find an umbrella to go into the street, but then I remembered that I'd left the umbrella at the well. Thus I lost my newly purchased umbrella and missed the eclipse.

If I was not reading and not writing, I was walking. Somehow, I was always in motion.

When I was not occupied, especially when taking strolls, I always sang. I did not sing songs, but sang like a bird without words. The words would have been further emanations of my thoughts, but I did not want this. I sang in the morning and evening, and this was a rest for my mind. The

[1] Pavel Mikhailovich Golubitzki (1845-1911), Russian inventor primarily with the telephone.

[2] Sophia Vasilyevna Kovalevskaya (1850-1891), Russian mathematician and inventor.

tunes evolved from my mood, and my mood was dependent on my feelings, impressions, nature and sometimes my readings. And right now I sing almost every day in the morning and before I go to bed, although by now my voice is already coarse and the melodies are all identical. I did not do this for anyone and no one ever actually heard me. I did this just for myself. It was kind of a need that I required. Unclear thoughts and sensations summoned sounds. I remember that this latent singing talent of mine surfaced when I was about 19 years of age.

In Moscow I had the opportunity to become associated with the well-known instructor Malinin.[1] I counted his textbook as excellent and I felt quite obliged to him. I spoke to him about dirigible design, but he told me, "This particular mathematician proved that a balloon cannot withstand the force of wind." It was useless to oppose him, since my evidence was essentially insignificant. But he soon died and so did Stoletov.

During one occasion in Borovsk, I was living at the edge of the city, and nearby was a river. Our street was unpopulated, covered with grass and very good for playing games. Once I saw at the neighbor's a small paper kite in the shape of a hawk—it was a Japanese toy made of reeds and cigarette paper. But it was damaged and did not fly. With the help of a pantograph I increased its size by several times, such that its wingspan was about a yard. I decorated the paper kite with pencil drawings and it flew beautifully. I could have even attached some small boxes to it. The string could not be seen and people from the ground actually thought it was a real live bird. It was a great illusion and especially when I would pull on the string and its wings would flutter, and it really looked like a flying bird.

What was impressive was how some large white birds flew near the toy and then realize it was artificial. Disappointed, they turned in another direction and flew away. Children and a crowd of adults walked to this place to watch. The movement of people to my end of the city bothered the town policeman and he became curious as to where the crowd was running. When he got near and saw not only the toy but also the string, in vexation he stated, "Can't these people get this into their heads, that this is not a genuine bird." Others thought that I had a tamed kite[2] tied to the string.

That night I released the kite with a lamp tied to it. The local residents saw what looked like a slow-moving meteor or planet Venus. Others laughed at me, calling me a fantasy-inventor. One person said I had lit a bird on fire and released it. At the time I was not entirely in my best of health and decided not to go chasing after my kite. But many considered this to be

[1] Aleksandr Feodorovich Malinin (1835-1888, Russian pedagogue: mathematics and physics.

[2] Here referring to the actual bird called a kite, similar to an eagle.

a joke. Anyway it compelled me to run to find it, and it seems that I again found my childhood. At the time I was about 30 years of age.

In Kaluga (ages 35–77, the years 1892–1934)

Here I started associating with the family of V.I. Assonov, who were well-known in the city, and then the family of P.P. Kenning. Assonov helped me get in touch with the Nizhni-Novgorod (today known as Gorki) circle of physics enthusiasts, whose president was the recently deceased S.V. Scherbakov. Initially with the help of the circle, and later independently, I began to publish my treatises on the sun, on flying machines, and others, in the journals: *Science and Life*, *Scientific Observation*, *News of Experimental Physics*, *Around the World*, and others.

The theoretical compositions of professors stated that there would be massive resistance, even with my best designs. Wanting to disprove them, I carried out many experiments regarding the resistance of air and water. I built the instruments myself: initially small and then larger, and which occupied almost all the space in my apartment. Occasionally I would lock the door, so no one would open it and not interfere with the air current that I was creating.

The postman is knocking, but I cannot open the door until the observations are concluded. The postman hears the monotonous sound of the metronome and me counting 15, 14, 15, 15, 14, and etc. Finally I open the door to the anxious postman. One relative, seeing the monstrous apparatus in my apartment, tells my wife, "When will he get rid of this devil?" One priest noticed that the holy corner was barricaded.[1]

Figures of various forms were glued from thick artist's paper. But on occasion what I needed for this was heavy wood planks. They were prepared for me by an engineer named Litvinov, who was a teacher. Never will I forget his unselfish services. He died and his son is now in Leningrad. We write to each other and I repeatedly thank his father for his efforts.

While still in Borovsk, a Moscow print shop offered to print my treatise *Aerostat*. To finance it, I provided half the money and the balance was funded by associates. An associate named Chertkov took charge of the publication. The booklets were published, but there was no financial gain for me from them. Even then the books poorly sold and the partners barely recovered their investment. Nonetheless, when I was in Kaluga, I received a copy of the booklet and I felt like I was in seventh heaven. What an unforgettable time!

While in Kaluga, a second volume of my *Aerostat* was published.

[1] Every Russian Orthodox home had a corner of the living room with a shelf and an icon of a saint on it.

By the way, regarding our children. All of them attended middle schools. All three daughters finished high school. The oldest daughter continued to higher education. The boys learned quite well in school, except for Vanya who was sickly from birth, although he did pass accounting classes.[1] One son died as a student. Another could not handle living in a metropolis; he passed the examinations as I did, to be a teacher in a higher institution, but he too soon died.[2] Now all I have left are the two daughters who live with me.[3] I have six grandchildren with me, the 7th is in Moscow with his father.[4] In the past he lived almost all the time with me, but now he is here only during the summers.

There were often music concerts in the city park, and it is amazing that I never missed even one concert. It was held in a pavilion. I was able to hear the music better than conversation, and often I would find myself singing along with the tunes in my squeaky voice.

I remember that after reading Well's *War of the Worlds*, all of a sudden a tune and words materialized within me, one that I never heard before, yet it corresponded with the destruction of humanity and complete hopelessness.

I continued my pre-occupation with electricity as always, now static electricity and galvanic action. I built machines of all sorts, and the most complex that I completed was an inductive motor with two rotors. The primary entertainment for my few associates consisted in my electrical achievements. They left quite satisfied, just like after a good meal. But now I have shortened my list of personal associates to zero, and only accept them on official matters or for the sake of scientific discussions. I can no longer tolerate mundane nonsense or shallow conversation or waste of time.

In 1897, I was assigned to give lessons at a state junior high school. But matters did not work out well for me, and then a new director arrived and he confiscated all of my lessons for his personal use.

At this time I went into depression. From my teaching position, I was transferred to the state junior high, and then I was moved to a third school in the county that was a big joke and all the young people did was fool around. There was nothing else for me, and with my poor health I was not able to endure what was occurring. I was then diagnosed with peritonitis. I thought I would die. This was the first time in my life I realized what it was to faint. While I was in despair for my life due to the pain, I lost consciousness. My wife got scared and started to cry for help, and then I regained consciousness, as though nothing happened, and asked, "Why are you crying?" She explained

1 He eventually died from natural causes.

2 These 2 boys died from suicide.

3 A 3rd daughter also died young.

4 This is the 4th of Tsiolkovsky's sons. They had a total of 7 children: 4 sons and 3 daughters. Only 2 daughters survived the parents.

to me what had happened and I realized that for some interval I had been in a state of non-existence.

In 1898, I had an offer to teach physics at the local women's diocesan school. I agreed, so after a year I departed from the county school and relocated. The lessons initially were few, but then I started to teach mathematics. I was teaching almost mature girls, and this was easier for me because the girls matured earlier than boys. I did well while teaching there and I did not have to fail any of the students.

On one occasion I gave a poor female student a grade of 5, but it was a mistake, but I decided to let it go. When she turned in her assignment the following day, she expected a 5. I noticed that giving unaccomplished students a high grade to motivate them actually reduced their motivation to study and was actually more harmful to them. Even though I was handicapped, matters went well for me while at this school. Teaching here lasted until 1905, when the school was closed. But matters still went well afterward.

Near my apartment and outside the city was a park. I would often walk there to think or rest, whether summer or winter. On one occasion I met an associate riding his bicycle. He offered me the chance to ride it. I tried riding the bicycle but failed, I just kept falling. Then I told myself, "No, never will I learn to ride on two wheels." The following year, 1902, I purchased an old bicycle and learned to ride it in two days. I was 45 years of age. Now I can celebrate 30 years of my ability to ride a bicycle. All of my children also learned to ride it, even the girls, except the eldest.

The bicycle was exceptionally beneficial to my health.

Grateful for this piece of mechanical equipment, I was able every day during the summer, during good weather, to ride outside the city to the forest. I also would go swimming. Although the Oka River was far from my apartment, it was close enough to ride to. In the past I would walk two miles to the school, but now riding made it easier. But seldom did I just ride about the city.

My finances for performing my experiments on the resistance of air were exhausted, and I turned to the president of the Physics-Chemistry Society, Professor Petrushevsky. He responded very kindly, but the funds for the Society were expended on publishing his textbooks. But the Academy of Sciences was able to help and contributed almost 470 rubles. I still have preserved the immense amount of data in tables with sketches. They were not published in the works of the Academy in part due to my obstinacy, but selections of my experiments did appear in many journals.

Meanwhile I continued to teach at the women's school. Thanks to the school official's continual surveillance of the facility, it was a very humane place and with a large number of students. Every class in the two departments, had about 100 persons. This place was more to my expectations. Due to my deafness I was still lacking in some areas in conducting classes. I tried to explain as much as possible so I would have as few questions as possible. But if a girl had a question, she had to stand next to me and talk into my left ear. These young voices were articulate and I was able to decipher their requests and evaluate their knowledge of my topic. Later I created a special ear trumpet, since they were not available for purchase at the time. The electric microphones were of especially poor quality and I did not use them.

At my convenience I would do chemistry experiments, like burning metals in oxygen. Burning hydrogen caused various whistling noises. I always showed the 5th grade students my dirigible. It would fly around the classroom on a thread and I would let students guide it. Such novelties as a flying globe, especially tying a doll to it, tended to capture the students' attention.

All the classes experimented with the effects of air pressure. I set up a glass bell jar under a vacuum stuck to a smooth plate. No matter how hard these young men tried, they could not lift the bell jar off the plate.

My steam engine would whistle.

My treatises continued to be published in journals, but no one took any note or considered them seriously. The only traces they left were in my soul and so I just continued higher and further. About this time I wrote and published my book, *Aerostats and Airplanes.*

My concepts of a jet propelled rocket was only noticed when it was published the second time, in the years 1911–1912, in a popular scientific journal. Then many educators and engineers—and those abroad—declared their precedence in this area. But they were not aware of my first treatise of 1903, on this topic, which proved my precedence.

During spring of 1914, before the war, I was invited to Petrograd to attend the aeronautic congress. I took a model of my dirigible that was about 6 feet long and made a report on it. Professor Zhukovsky opposed my report and did not approve of the project. His students continue to this time to hinder my progress. But suppose they are right; I will not believe them until they show it to me.

The students who watched my presentation said that it was only because of the model that they clearly grasped the concept of this new type of dirigible. My books just did not present it properly. This is how difficult it is to adopt something that is new.

We were glad for the revolution. We hoped for an early end to the war, and some freedom. Due to my age, I conducted myself rather apprehensively, not attributing significance to slogans, and not once did I even wear a red scarf or banner. So what occurred at one school where I was teaching, they thought I was a reactionary. But I showed them a book that I had published when the tsar was still in power, that was purely socialist in content. Now I was definitely a Bolshevik. I did not like to publicly display my attitude toward the revolution.

The schools were reformed with the October Revolution. Grades and examinations were banished, and equal rations for everybody and a general right to work were introduced. In short, they introduced the most idyllic communist rules. In Moscow the Socialist Academy was instituted. I introduced myself to them and sent them a resume, and so was selected a member. But I was already declining in health, in addition to being deaf, and I could not fulfill the requirements of the Academy and relocate to Moscow. So after a year passed I had to resign from the Academy. I retired from work in 1920, and entirely terminated my teaching career. I did receive an academic ration, and then help from the Central Committee to Improve the Conditions of Scholars[1] in the form of a pension that I continue to receive to this day.

But I did not abandon my work. On the contrary, often I worked so hard and so much, more than ever, after I left my teaching career. The total time of my teaching career spanned 40 years. During this course, I managed to educate at least 1,500 students who finished middle school, and about 500 students who finished a higher education.

I also dedicated myself in later years to writing treatises on socialist concepts and natural philosophy. Some of them were published, while the majority are presently just in manuscript form.

The basis of my natural philosophy starts with a complete rejection of the routine present comprehension of the universe that is provided by contemporary science. The science of the future, of course, will succeed the science of the present, but for the time being, contemporary science is the more respected and even the sole source of philosophy. Science, observation, experiment, and mathematics, are the basis of my philosophy.

All preconceived ideas and teachings were discarded from my comprehension, and I began from the start all over again, and utilizing natural sciences and mathematics. The sole ecumenical science of objects and matter was the foundation of my philosophic thought. Astronomy, it seems, played the primary role, since it provided a wide periphery. Not only Earth's displays were material for my conclusions, but also the cosmic: all of these innumerable suns and planets. It is Earth's displays, the imperfection of Earth and humanity as the result of a still infantile stage of growth, that led almost all thinkers into deception and pessimism.

1 Центральная комиссия по улучшению быта ученых (1921–1937).

Now that I have a pension from the Soviet government, I can freely dedicate myself to my efforts and pay attention to my work without worrying about superficial responsibilities. My dirigible was recognized as an especially important invention. To further my research on jet propulsion, the Group for the Study of Jet Propulsion[1] was established as an institute. Many treatises in newspapers and journals were published pertaining to my achievements and efforts. A celebration was also held in Kaluga and Moscow on my behalf. I was awarded the Order of the Red Banner of Labor and a medal of honor from the Society of the Cooperation of Defense, and this increased my pension.

The USSR is progressing mightily, driven along the great path of communism and the industrialization of the country, and I cannot but be deeply touched by this.

A Traveler to the Spatial Expanses

(1928)

Seventy years have passed since my birthday. I worked the entire time. Now my strength is declining, I am in pain, but yet I still work.

I was born in the village Izhevskoye, Ryazan Province. My father served in the forestry division as a official, and he received a small compensation. We were not wealthy and his income did not stretch far. I was only able to receive a domestic elementary education. In 1880, I successfully passed an examination to teach mathematics and some sciences, and from that time until I completely retired in 1920, I taught in secondary and special educational institutions, and I likewise taught mathematics and physics at a national university, living for the most part in Borovsk and Kaluga. I taught mathematics and physics to over 2,000 young people. I became a member of the Society of Amateur Space Explorers, the Astronomical Society in Kharkov, and others.

And now I will explain how I lived and the purpose of my life. At about the ages of 8 or 9, I first saw a toy aerostat and I became interested in it, and as a result I started to build my own small aerostats out of paper. Then I started to build a carriage that moved due to the force of wind upon it, making it self-propelled. I would skip breakfasts and use the money to buy

[1] Группа изучения реактивного движения (GIRD)

nails. But this experiment was not crowned with success: in part due to my impatience and the lack of materials, and in part due to being tired of hunger.

Beginning at age 15, I was seriously attracted to aerostatics, and since I already knew the mathematics and a sufficient amount of data in order to resolve the questions of its design, and with a metal shell, I was able to determine the necessary size of a balloon and its length so it would rise into the atmosphere and carry some number of people.

The concept of centrifugal force also interested me and because I felt I could apply it to an invention that would rise as high as the cosmic spatial expanses. There came a point in my life at the age of 16 years that I thought I resolved this question, and I was so excited, so overwhelmed, that I wandered Moscow the entire night and thought about nothing else except the great results of my discovery. But by morning I was convinced of the fallacy of my invention.

The concept of communication with other worlds of outer space never left me ever. In 1895, I first explained, although cautiously, my various reflections of this concept when I composed my *Dreams of Heaven and Earth*, and finally it was defined completely in detail in my later composition *Exploration of Outer Space Worlds using Jet Propelled Rockets*, which was published in 1903, in the magazine, *Scientific Review*.

Today I can state with satisfaction that I possess many followers among the scientific class, in the USSR as well as in other countries. These are for example: Prof. M. Wolf, Prof. M. Weber, Prof. H. Oberth, Prof. R. Goddard, Prof. L. Schiller, director V. Goman, director M. Valye, Mr. Laren, Mr. Chander, Mr. A. Shershevski, and others. These scholars continue work that resulted in the theoretical and practical affirmation of my ideas.

Recently I likewise published several compositions where I endeavored to develop and confirm with the help of mathematical analysis the basic principles of my inventions for interplanetary travel. These were: 'Rockets in Outer Space" (1924), "Outer Space Research Using Jet Propulsion (1926), and 'The Cosmic Rocket." As a young man, many other questions agitated me and drove me to undertake difficult experimental tasks.

So when I was at the age of about 23 or 24, I presented a series of papers at the Petersburg Physics-Chemistry Society. These were: "The Theory of Gases," "The Mechanics of a Living Organism," and "How Long the Sun Will Continue to Radiate." Professors Borgman, Mendeleev,[1] Van der Flit, Sechenov, Petrushevsky, and others, rated my papers highly, and this overjoyed me. I contributed additional papers to the Astronomy Society and they published them: *'Gravity as the Source of Universal Energy*" and "*How*

[1] Dmitri Ivanovich Mendeleev, the famous Russian chemist and inventor, and developer of the Periodic Table of Elements.

Long the Stars Will Continue to Radiate." But about two or three years after their publication, the famous astronomer See[1] claimed he had already come to the same conclusions long before I did.

From 1885, I firmly decided to devote myself to the study of airships and to develop a design for a horizontal metallic aerostat with conical ends, and which was navigable. I worked on this project almost 2 years without stopping. Finally in 1887, I made my first public presentation in Moscow of a metallic, navigable aerostat. My presentation interested professors Veinberg, Michelson, Stoletov and Zhukovsky, but the project did not progress any further. Then in 1890, I turned to D.I. Mendeleev, with a letter and eager to work, but asking him for his opinion on my metallic aerostat. Mr. Mendeleev responded that he also at some time in the past spent time on this question, but then discarded the project. But he promised me he would pass my manuscript and model to E.V. Feodorov of the Technical Society. Mr. Feodorov checked my calculations and reviewed my report and responded that the concept of building an aerostat from metal did deserve some attention, since metal will not release gas as it is non-porous, and this would allow travel to be less expensive and provide the ability to conduct long flights. But even with his compliment and assurance, the matter just died, and as a result of the opinion of other recognized specialists of the era who claimed that an aerostat would always be nothing more than a toy subject to the whim of turbulent wind.

Then in 1892, I published a book titled *The Metallic, Navigable Aerostat*. This book passed though Russia unnoticed and unconsidered, but in 1897, an article in the *Paris Scientific Review* indicated a parallel between my work and the work of the famous Andree, who perished at the north pole with his dirigible.[2]

It was only after many attempts that I was later successful in providing a final design on the possibility of the construction of a metallic dirigible with conical ends. My project is presently beginning its materialization in the USSR and, judging by what is printed in newspapers, it is also gaining some interest abroad. A positive response was also provide by Roald Amundsen. I am satisfied that my effort was not futile. I retain an optimistic view of the future of humanity.

I believe that humanity will not only inherit the earth, but will also inherit the region of planets and perhaps the region of stars. This thought I have been developing in all of my special efforts, for example, in my book, *Beyond Earth* (1920). The population of the universe by humanity from Earth will absolutely happen, and as soon as Earth becomes too crowded, and

[1] Thomas Jefferson Jackson See. He was considered a fraud by many American astronomers and scientists.

[2] S.A. Andree balloon expedition to the North Pole of 1897.

technology will considerably increase to the point that the aspiration of a person will expand beyond his surroundings, and to advance will be easily achievable.

I was successful in theoretically proving that the technology of the future will provide the possibility to overcome Earth's gravitational force and we will be able to visit and learn about and teach all planets. The imperfect worlds will be liquidated and replaced with an advanced civilization. Resources to accomplish this will be provided by materials from asteroids, planets and their satellites. This will allow the inhabitants to flourish, and even to the point of increasing universal population to a billion times more that Earth's population. As a result the universal population could reach to a billion billions of entities.

At the present time, I have finished my new research paper, "Air Resistance and a High Speed Train" (1928). On the basis of purely theoretical calculations, I came to the conviction that another completely new method of propulsion can be applied in the future for faster movement. My super-express presents a vehicle of a known design with a smooth bottom, without wheels, and will ride on a cushion of air. This will allow the velocity of the vehicle to reach 700 miles per hour. My calculations even claim that such a vehicle will cross the widest rivers without the need of bridges by practically flying over, even if several miles in width. But this project of mine needs to be approached as still a technical task that we cannot resolve in just one day, meaning today, but will extend into tomorrow, the future.

Such an existence can be acquired, always staying somewhat ahead of time and its possibilities—those within which we reside, always considering objections to the progress of thought, to foresee disbelief, to react against importunity, and not incur losses. Our sphere is the sphere of speculation, to hypothesize. But how are we to walk along a untapped and unknown and still dark route of a scientific-technical concept, unless we illuminate this route with the light of science fiction?

Fate, Destiny, Fatalism

(July 1919)

All my life I complained about my destiny, my misfortunes, the obstacles I incurred attempting to have productive accomplishments. Where they accidental, or did they possess some kind of meaning? Did they lead me along some assigned route with a high assigned purpose? I will here strive to answer this question.

I was born in 1857. During the years 1867-68, when I was about 10 or 11 years of age, destiny dealt its first stroke at me. I developed scarlet fever and the result of which was some mental retardation and deafness. Up to this time I was a happy and capable child, and they loved me a lot, always kissing me, gifts me toys, candy, money.

What would have occurred with me if I did not lose my hearing? If course it is impossible to foresee this, but the following would have more than likely occurred: due to my inherent abilities, health, cheerful character, talent, I would have proceeded along a blazing trail. I could have concluded my classes earlier, finished some civil service, acquired a vocation, married and had a lot of children, accumulated a lot of property and possessions, and died content and satisfied with my limited life, while surrounded with many descendants and devoted people. My mental faculties would have always been rested, I would have been calm and prosperous, and very gifted by nature.

Perhaps I would have accomplished something small, wrote some kind of book, developed some kind of philosophy, perhaps collected some royalties and existed. But all of this would have been very insignificant and doubtful, since happiness and satisfaction tend to stifle higher level activities.

So what did my deafness cause for me? It caused me to suffer every minute of my life, especially when with people. I sensed myself always isolated when with them, deprived and derelict. This caused me to look deep inside myself, forced me to seek greater matters in order to earn the people's approval and so I would not be so disdained. It always seemed to me that I was disdained due to my deafness. And it was this way, and in the same way we tend to hide our disdain for the retarded and crippled even though we possess it. The initial impact of my deafness was an increase in my mental faculties, which caused me to care less about the impression people had of me. It was as though I became numb or callous to them, oblivious to the constant ridicule or offensive remarks. As a result my capabilities weakened; it was as though I submerged myself into darkness.

I could not learn in school. I could not at all hear the teachers or even hear any clear sounds. But gradually my intellect found another source of ideas: books. My interest in physics, chemistry, astronomy, mathematics, and other subjects, developed when I was 14 or 15 years of age. Of course books were few and I submerged myself more into my individual contemplations. I did not stop thinking about whatever it was I read. Much I did not understand and there was no one to explain anything to me and even then with my handicap this would not have helped any. More than anything else, this motivated my intellect to function on an independent basis. The deafness caused my ambitions to suffer uninterruptedly, it urged me forward, it was a knot that pursued me all my life and still now pursues me. It caused me to separate from worthless people, from any routine or mundane pleasures. It forced me to concentrate, to devote myself and my entire intellectual faculties to science. Without it I would never have accomplished or completed so much work.

If this had occurred to me 10 years earlier, then I would not have learned the language and grammar, and had but a rudimentary education, but sufficient health and ability to deal with life. I would not have been able to progress to science, not tolerate the impact of deafness on my life, and I would easily have become deaf and even illiterate.

If this would have occurred much later in my life, I would not have been able to deal with the grief, I would not have accustomed myself soon enough to contemplate, I would become infected by various people's ideas and attached myself to any of them.

But it was not deafness alone that made me turn out to be what I am. I did inherit redeeming and beneficial qualities, however what followed was a series of jolts and cruel blows that finished what my deafness started on my character development.

My father did not have much financial means. At about the age of 21, when I was relieved from military obligations, he wanted to buy a house and there father and I and my sister and aunt would live together. This was in 1878, and they hoped this would determine my destiny. But here again an insignificant jolt occurred that directed my life's path along a completely different route. My father had a microscope. I looked at something through it and rotated it, and I lost one of its lenses and was not able to find it at all. The microscope was antique and simple. My father did not like this type of sloppy work, and so I was afraid of his displeasure and utilized all my effort to find the tiny lens, but I still did not find it and did not tell father, either. Shortly after, he lent this instrument out and received it back. Looking closely at it, he noticed it was missing one of the oculars. He became angry and complained to me, "This is what happens when you lend somebody something."

Then I told him that I had lost the lens. He became terribly angered at himself because he had already blamed his innocent friend, and then he was angry at me. An unpleasant scene occurred and all because of me. The result was a rupture in the relationship between my father and me. This occurred in about October or November 1879, the following year. Then I left home and got my own room, but I had nothing to live on. It did not take me long to exhaust what savings I had for my survival. I had to come down from the clouds and study the topic I hated most: Orthodox catechism, in order to pass an examination to become a mathematics teacher. I think it was September 1880 when I passed the examination and received a diploma as a mathematics teacher in Borovsk, Kaluga Province.

This extremely minor incident of losing a lens completely changed my life. I was completely occupied with myself. I made my endeavors my first priority. I was chock-full of other-worldly ideas, or more correctly said, those that were alien or transcendent to most people: I was eternally soaring in the clouds. At the time, I also studied the Gospels passionately.

In those days I had a very altruistic character, a happy public image. I was drawn to women, I fell in love with one after another after another, although none of this interfered with me preserving my immaculate, unblemished chastity, not once did I violate my prudence. I was likewise faithful my entire marriage. I decided to not follow my initial intent but rather to marry as soon as possible. It was a loveless marriage but to a good and industrious girl who found my projects irresistible. The autumn of that year I accomplished this objective. Then my father died (I had long since reconciled with him).

My ideas consumed and overwhelmed me and drove me onward to such an extent that when I left Ryazan, I never again saw any of my family or relatives, although I did correspond with them.

Marriage was likewise my destiny and a great motive force. Perhaps I can describe it as attaching awesome fetters to myself. I was tricked by life: my children were angels, just like my wife. The domestic instinct was gratifying enough, but some greater passion forced my intellect and strength to strain and seek. It seems that an uninterruptedly motivating and insatiable heartfelt sensation was appended to the eternal humiliation of my deafness. These two forces pursued me in life, and pursued me in a way worse than anything that entered my mind, or was impressed on me, or taught to me. I was continually seeking, seeking independently, traversing from these difficult and serious questions to others that were even more difficult and important. It was only science that restrained my thoughts and fantasies.

But books were few and I had no teachers at all, and so I had no choice except to create and generate rather than accept and adopt. Direction and assistance did not exist anywhere, there was much in books that I did not understand, and so I had to explain it all to myself in the best manner I could and come to the right conclusion and not err. In short, the creative element, the element of self-development and originality was controlling me. I can describe it as a life-long non-stop state of learning to contemplate, defeat obstacles, resolve question and tasks, all on my own.

Many sciences I straightway and independently created on my own due to a lack of books and teachers.

What happened was that my friends seemingly all of a sudden just scattered. I sensed myself distant from amiability. I strove to suppress the lonesomeness with new hobbies. I felt so lonely, but I never drank. Not once did I ever get drunk. I started to slide into despair. Earlier I was enthused by the Gospels. I attributed an immense significance to Christ, although I never ranked him among the deities. I saw a destiny also being worked in my own life, higher powers were guiding me. From a purely materialistic view of things there interfered something mysterious, belief in some kind of unfathomable powers tied with Christ and the First Cause. I thirsted for this clandestine concept. It seemed to me that it could restrain me from despair and give me energy. I wanted proof of this and to see it in the clouds in the form of a plain figure, a cross or a human. I wanted this and thought about it and then forgot about it.

I guess a few months or maybe several weeks passed, and we moved to a different apartment, the Kovilyov house, Kaluzhski Street, the house that burned down within a year, in the city Borovsk. I became interested in the direction of the wind, weather and such. Suddenly I see toward the south, and not too far above the horizon, a cloud in the form of a cross with each of the four arms identical in length. The depiction was so accurate that I was awesomely surprised and loudly shouted to my wife to come and watch his strange cloud. Of course she was occupied in the house with something, so either she did not hear or just did not come. When I watched it, I completely forgot about all my own aspirations. For a long while I followed the cloud and its form did not change any. This started to annoy me and I started to look at other things or maybe became absent-minded, I do not know right now. Then I waited for a while and again looked in that direction. Now I was no less amazed, as the same cloud reconfigured into the shape of a human being. The figure was at a distance and medium-sized, but its hands, feet, torso and head were distinctly ascertainable. The figure was also straight, flawless as though coarsely cut out of a sheet of white paper. Again I called my wife, but again she did not arrive. The figure remained the entire time while I was looking at it. I could have run and called my wife, but the scene seemed to me to be so interesting that I selfishly could not tear myself from it.

Then I remembered that I had contemplated having a vision of such figures earlier. Neither before nor later in my entire life did I ever see anything similar. I guess this occurred in the spring of 1885, when I was 28 years of age.

It was a wretched time for me.

This strange apparition was tied to my preceding thoughts and moods and had an immense influence on my entire life. I always remembered that there was something inexplicable, that the Galilean teacher presently lives and has a significance and imparts his influence to the present time. This sparked interest deciphering my difficult life and encouraged me. I told myself that all was still not lost, that there is still something out there that can support me, maintain me.

Disregarding that I was saturated with contemporary viewpoints, a pure scientific spirit, simultaneously something else incomprehensible stirred me and uneasily stimulated me. This was the cognizance of the lack of comprehensiveness of science, the possibility of error and humanity's limitation, its distance from the true orientation of things. It remained with me and now even grows with the years.

In 1902, a new stroke of fate jolted me: the tragic death of my son Ignaty. Again I had to deal with a horribly wretched and difficult period in my life. From early morning, as soon as I would awaken, I already felt empty and regretful. It was only after 10 years that this feeling dispersed. This sorrow and the accompanying sensation—the thought of the hopeless despair of people who have lost the ground under their feet and the desire to live due to loss of a child—compelled me to write my treatise *Ethics*, at the beginning of 1903.

This misfortune softened my heart, tamed at least some of my character, directed me to heaven, to the future, to infinity, and perhaps delivered me from the possibility of doing something bad or wrong. If it was not for this grief, I would not have written my *Ethics*. The death of one saved many, and I do not think that his life was unproductive, or his death.

Right from my childhood, the character I possessed was despicable, burning and unrestrained. But due to my deafness—destitution, humiliation, and emotional dissatisfaction materialized, and along with it was a fiery obsession to the point of insanity of my push for truth, science, the welfare of humanity, the aspiration to be useful, a means to terminate my self-consciousness, however the resultant course ended up being the complete neglect of regular human responsibilities.

As least important I placed the welfare of my family and associates. Everything was done for a higher goal. I did not drink, did not smoke, did not squander even one extra kopek on myself, on clothes. I was almost always half-starving, poorly attired. I moderated myself in every manner to the last degree. My family suffered with me. I have to admit that they were sufficiently fed, warmly clothed, and always had a warm residence, and did not eat just simple dishes, but we had plenty of wood and clothes.

But often I was irritated at myself for these poor conditions and perhaps for making life hard and rough on those surrounding me. I did not have a heartfelt connection to my family, but I had affection toward them, but one that was more theoretical and not natural. And this was hardly easy on the people who surrounded me. I was compassionate and honest, but I did not possess a simple or passionate human form of love.

Cosmic Philosophy

The Will of the Universe

It is as though all depends on the will of some intelligent entity, similar to a human. Our effort and thought gains the victory over nature and directs it along our desired channel. For example, we cultivate land and receive an abundant harvest; we domesticate and train animals; we improve plants; we build homes, roads, machines that make work easier for us; we utilize the forces of nature to work for us; and they increase our strengths by 10, 100, 1000 times. If we were lazy and did not materialize our will, then nothing would exist and we would perish from starvation, cold, illness, infertility, and etc.

This is conditional will and it contains a beneficial materialization. Right now it is still not great, but limited, but we can hope that it will grow and disclose itself as stronger. Conditional will is the incarnation of our thoughts and desires in life. For example, I want to build a house, and while building it I need to buy some kind of machine, and I buy it in order to fulfill my wish. I want to marry some woman and I do marry her. I want to have children and I have them. In the majority of situations our intentions are not possible, especially the difficult ones, but theoretically we can admit that a strong will does exist and which can cause all of our reasonable wishes to be fulfilled. If not now, then in the future, and if it is not us, then it will be accomplished by other more capable persons or even by our descendants. There is nothing more transcendent than a strong and intelligent will. Just to have intellect without a will is useless, and to just have will without intellect is likewise redundant. Every entity must live and think as though he can attain sooner or later all of what he wants.

But what is the source of this beneficial will? The will depends on the orientation of the brain. The higher brain evolved from the development of the brain of the lower animals. All animals and plants evolved from complex organic matter, such are bacteria, amoeba, protoplasm, and which themselves evolved from inorganic matter. The development of the organic world is impossible without some kind of energy, for example, the energy of the sun.

It is clear that nature gradually generated life, intellect and will. A person is born of soil; the soil evolved from the sun; the sun evolved from the concentration and precipitation of dispersed masses of various gases, and they evolved from even more dispersed matter, for example, from the ether.

So everything is generated by the universe. It is the origin of all created matter and all depends on it. The human or any other higher creature and its will is only the manifestation of the will of the universe.

Not one creature or entity can unveil absolute will, just as a clock cannot unveil it or some other kind of complex automatic equipment, for example, talking movies.[1] The movie characters talk, walk, act, do what they want, and their words correlate with their actions, but every person knows that their will is only seeming, it is not absolute, all their movements and talking depends on the film, or on the person who created this movie. So it is that the intelligent entity fulfills only the will of the universe, since it gave the entity intellect and a limited will. It is limited because this will depends on the extent of the intellect and cannot be the sole source of its behavior: the bulk of the universe can always intervene, and mutilate and alter, and so cause the will of one intellect not to materialize. We say, "It all depends on us," but we ourselves are just a creation of the universe. For this reason it is more proper to say and think that it all depends on the universe. We have in mind to do something, but the universe makes arrangements as it wants, without ceremony it destroys our plans and even destroys the entire planet with all of its intellectual creatures.

If it should be possible for us to fulfill our will, then this is because the universe is permitting us to do so. It always possesses a quantity of abilities and reasons to put the brakes on our intentions and insert some other higher will, even though our will is only the will of the universe.

Let us take the events of this year, 1928. It is clear that they are the results of the events of the year 1927, and the latter are the results of the events of the year 1926, and etc. In the final end, whatever happens has the eras long past as its source, even long before there were any traces of roaming animals. Our will, our actions, those of the present and the future, are the

[1] This article was composed in 1928, and talking movies just started to appear in Soviet Russia.

result of long past eras. And these were generated from times that are even earlier. The extent of time into the past is a decillion years,[1] or a decillion times a decillion years, or a decillion to the decillionth power of years, all into the past, and these provide the conditions of the world and its reasons for contemporary and future occurrences.

What interests us is not so much the display of a person's will—although it is generated by the cosmos, as much as the general display of the will of the universe.

Do we have the right to speak of the will of the universe? We can speak of the will of an intellectual creature, the will of a dumb animal, but can we speak of the will of the cosmos? Can we compare it even to a dumb animal? But since all depends on the arrangement of the universe—meaning at the present or long passed moment—this means it or its unknown cause possesses a will. This will provides conditions for all else that we presently see or to what our intellect directs us. The question now is focused on: Whose will is this? The quality of the will indicates to us the quality of the universe or its causes.

When we speak of the contemporary state of Earth, the will of the cosmos is particularly displayed as the will of the dumb creature. Indeed, in the matters pertaining to Earth, in the matters of humanity, we see a mixture of the intellectual with the stupid, the fine with the cruel. So why do poverty, illness, prisons, malice, war, death, stupidity, ignorance, the limitation of science, earthquakes, hurricanes, poor harvests, droughts, floods, harmful insects and wild animals, bad climate, and etc, exist? Just looking at Earth, we should compare the will of the cosmos with the will of limited creatures. But will Earth and humanity always exist in this condition? The universe generated humanity, its weak mind and will. Although they were even in earlier times, they then developed to the strength they now possess and there is hope that they will continue their development. What level they will attain, what they will produce, this is difficult for us at present to imagine. (This topic is further developed in my book: *The Future Earth and Humanity*.)

So a complete assurance exists in the possibility that the will of the cosmos will be displayed on Earth in all its brilliance of a supreme intellect. The perfected condition of Earth will continue for a very long time in comparison to its sorrowful state, as at present. Our posterity will only be able to blame this will as a short decline during this interval: why the cosmos did not immediately create happiness but forced part of its composition to first experience disorder, suffering and fear. Although such sufferings are comparatively short, however they did need to occur. Not in any circumstance do they compose more than a thousandth of the time of

[1] 10 to the 33rd power

the future prosperous state of people. We will then watch such blame just fall to the wayside.

There are over a thousand planets in our solar system and only one of them possesses the conditions that are conducive for the development of higher conscious existence. In our Milky Way, there are over a billion solar systems similar to ours. In our Ethereal Island[1] there exists over a billion Milky Ways. We could continue this infinitely, but will limit ourselves to our Ethereal Island, where were can quantify millions of billions of solar systems and at least the same number of planets capable and conducive for the development of organic life. Life can be generated on all planets, but on these it should be able to flourish. The question arises: Were all of these millions of billions of planets the same? Without doubt, evolution was earlier on some of them than on others, and some completely extra-ordinary due to their beautiful results. Life was assigned there to have more possibilities, and all the best, and which a person at present can hardly imagine. Such a person dreams, not only to prevail over his solar system, but to visit others. The goal is to liquidate all his congenital imperfections painlessly and populate the planets with his perfected posterity.

The person of the future will think, "Organic life on Earth has passed such a difficult course, and its residents and inhabitants have endured so much suffering. Now they are no longer here. Other than the intelligent, nothing populates our planet. So what was this terror of the past? Would it not have been better if the Earth was from the start populated by conscientious creatures. Then this tail of suffering and insanity would not exist."

If a human can think in this manner, then how much more would a higher creature on a higher planet think. And not only think, but would find thousands of possibilities to materialize his intelligent intentions.

So the intellect and might of the higher entities that are generated on higher planets will liquidate the rudimentary or embryonic life on other planets and populate them with their posterity. Similar to this is the gardener who uproots weeds in his garden and then plants useful vegetables there.

What will turn out is that the universe will be filled with only perfected entities. On the infinite majority of planets a perfected life will directly arise and stand on its feet, without the long millennia of preliminary pains of childbirth. From the millions of billions of planets, only a few have endured the cup of suffering, and so life on them begins with the pains of childbirth.

If you during the course of your entire prosperous life are tormented only one second, then would you consider your life to be unsuccessful? So you should not consider the universe to be unsuccessful on the basis that some

[1] *Ethereal Island* is Tsiolkovsky's term for the local universe with which we identify.

planet from the billions of them must endure a comparatively infinitesimal time of pain in childbirth. If the cosmos in general provided only happiness for its entities, then it is possible to consider its will as irreproachable.

The cosmos did not generate evil and error, but intellect and happiness for all its creation. In order to understand this, it is necessary to only stand on the higher point of vision: imagine the future Earth and use intellect to embrace the infinity of the universe or at least just our Milky Way. Then we will see that the cosmos is similar to the best and most intellectual entity.

You perhaps will say, "Well now, happiness and intellect are flowing everywhere, but are there not elements of disaster that can in just one moment sweep away this happiness, like a broom sweeps away useless trash? The planets and suns can explode as do bombs. What life can survive this? The lamp is extinguished and deprives the planets of its life-giving light. Then where will its inhabitants go? Multitudes of catastrophes always seem to entrap intellectual creatures."

The matter is that we possess a very shallow and unreliable understanding of the intellect of the higher creatures. If people should foresee certain disasters and take measures against them—and they do often successfully deal with them—then how great the strength of opposition that the higher entities of the universe possess to also deal with them. They foresee the rupture of planets hundreds of years in advance of its occurrence and depart from them into safe regions of the cosmos. They foresee the explosion of suns, and likewise their extinguishments, and so depart at an opportune time from the weakening ones.

You will also say, "Sooner or later all the suns will extinguish, life will terminate. So here is your beneficial cosmos." But this cannot be. Suns uninterruptedly re-ignite, and this occurs more often than they extinguish. In short, the dark suns resurrect no less often than the bright ones extinguish. The universe always was and will be viewed for the most part in the manner that we observe it today. The life of every lamp is periodical and repeats a multitude of times. So does the life of groups of suns, for example, the Milky Way.

"But death and the agony that it includes," you will say, "can this ever vanish?"

Life does not possess an indefinite extent, and it can be prolonged up to a thousand years. Death can be painless, like the death of a tree or some kind of insect. Other than this, I have proven many times that the life of a lamp is periodical and turns on and off an innumerable number of times, and so does the life of an entity, or the atoms that compose its being, resurrect many times, or more correctly said, they always sprout and will sprout without end.

Not one atom of the universe will escape the effect of the higher intellectual life. Besides this, only such a life is possible.

Death is one of the illusions of a weak human mentality. It does not exist, because the existence of an atom in an inorganic material possesses no memory and time, and as if time does not even exist. The majority of the atoms existing in organic form are blended into one subjective continuous and happy life, and a happy one, since there is no other and which is not allowed by the intellect and powers of the higher animals.

The universe arranged so that it is more than just immortal; all of its parts are immortal in the form of living and blessed entities. There is no beginning or end to the universe, and likewise, no beginning or end to life and its bliss. We demonstrate that the will of the universe is beautiful, because in the general picture of the cosmos we see nothing except beauty, goodness, intelligence, perfection and their subjective consistency, beginninglessness and endlessness.

If the cosmos has a cause for existence, then we must also ascribe the same qualities of general love to this cause.

We are assured that mature entities of the universe possess the means of migrating from planet to planet, to intervene in the life of the balance of planets and communicate with the same type of mature entities. People of Earth will at some time unite and all of them will govern through one elected council under the guidance of a president who is selected by this council.

This will occur relatively soon. Within a more significant interval of time, our entire solar system will be abundantly populated. It will also be governed by an elected council with its president. It cannot be otherwise, since this is what requires intellect. Likewise all the other planets and solar systems will unite. The difference is only that the majority of them are quickly populated with mature and already perfected posterities and immediately install their government. For our Earth and our solar system, there has been a prolonged route of gradual development through martyrdom. There must be a unification because this is needed for the advantage of the created entities. If they are mature, then they are intelligent, and if intelligent, then they will not cause damage or do harm to each other. Anarchy is imperfection and malicious.

The presidents of groups of suns (collections of stars) of the entire Milky Way will unite. This is the basis of their future.

So what type of organization does this have to be, in order to be mighty as well as wise? We speak of entities that are similar to humans, but more perfected. Different races or breeds can be among them and those that are adapted to life on every planet, and Earth, for example. However the majority of them are identical and oriented for life in the ether. But others are also needed who can implement order on all the planets. This order consists in banishing all suffering in the heavenly bodies. This means that we can expect this mighty organization to intervene into any planet, and Earth for example.

Why is it that we do not notice any traces of such activities to this time? Science is precise, and not one branch of the mind is distinguished to this extent by such sobriety, yet our imagination is weak. It has deceived people so many time, and at the present time its credibility has strongly collapsed. The sobriety of science has not permitted interplanetary communication to this time. Now this opinion is unsteady even for the scholarly, but the majority of them are still not enveloped with new ideas and possess either an indifferent or malicious attitude toward them. Somewhat earlier, and other than for the fantasy visionaries, no one promoted the possibility of spatial travel, special excursions outside Earth. This is why the opinion that this was impossible was concluded. And if it is this way, then all the facts proving the existence of such communication, if they did exist, were ruthlessly rejected by people of science. They also rejected the concept of meteors, objects falling from space to Earth, and for a long while in history, no sunspots were ever seen. So is the force of prejudice.

Meanwhile, history and literature note a number of unexplainable apparitions. The majority of them, without doubt, can pertain to hallucinations and some other class of delusion. But are they all? Now in view of the evidenced possibility of interplanetary communication can we approach such unexplainable apparitions more attentively. But we still lack concrete evidence in this area and so need to set aside the topic for a while.

Perhaps the development of the majority of people has not yet developed them for the intervention of other entities into Earth's life. But perhaps, it would even harm humanity at the present time. The majority of people look at the universe in a completely ignorant manner, just as do animals. Their religious views are basically superstition. If they were to see the intervention of other entities in Earth's matters, they would immediately perceive it from the point of view of their religion: Fanaticism would surface and nothing more.

Suppose the powers of other worlds would have stopped the war of 1914. There would not have been a war. Many people would have been delivered from suffering and death. But humanity is so coarse that it takes suffering such as this to motivate a repulsion toward war in them. What can be done with people such as these, that only horrible sufferings can mold their thinking to lead them to do better? And this is why it seems that a criminal cannot grasp his situation unless he is faced with some cruel incarceration. A rowdy drunk needs to be tied, while an insane person needs a straight-jacket.

There is another series of apparitions, and truly in the majority of incidences they are just as doubtful as those mentioned earlier. They speak to us in mental influences that effect our brain and intervene in human affairs.

I myself twice in my life was a witness to such apparitions and so I cannot deny them. And if they occurred to me, then why can they not occur to others? Based on this, I need to admit that some portions of such types of apparitions are not illusions, but definite proof of the existence in the cosmos of unknown intelligent powers, some kind of entities that are not built as we are, but are at least built of a more subtle matter. Perhaps, this legacy of a decillion years of progress of previous eras created such entities, as apparent in these present apparitions. They likewise display themselves in beneficial efforts. This is understandable because the mind and intent of all entities of the cosmos of all the era consists in nothing being permanently imperfect, and not having any suffering. And if there does exist suffering, then not even a atom will escape it. If there is no suffering, then not even an atom will suffer.

The past has no limits. What does this mean? Do not think that this is easy to imagine. I can provide an example. Here are some numbers: 22 meaning 2 to the 2nd power, 333 meaning 3 to the 33rd power, and etc, or conditionally X to the (x) power. If (x) is equal to 2, then this number becomes 4; if it is 3, then the number we conclude is about 5 quadrillion.[1] If x=4, then the number that is expressed will have so many zeros behind it that there is not enough room in the known universe to fit them.[2]

In reality, the volume of the entire known universe consists of the Ethereal Island containing a million Milky Ways over the expanse of 100 million light years. Suppose (x) is a decillion, then our function to express the number is still small relative to the factual infinite extent of the universe. So how large is it really and relative to time that has already progressed in the past? Our ability to fathom what we see and know is almost a zero amount compared to its true size.

So what can these infinite eras of progressed time produce? If time in a short interval is able to generate intellect and other miracles of life, than how much more can it create in the defined time of (X(x)), using the number decillion?

The world has always existed. The present matter and its atoms are an immeasurably complex product of another more simple form of matter. There were past eras when matter was a decillion times lighter than what is today the lightest. There were epochs when it was a decillion times decillion

1 Which is 5 with 16 zeros following it.

2 Which is 4 to the 444th power.

times lighter. And all of these worlds existed and generated intelligent entities, but almost intangible due to their low density. What I say, I have long theoretically supposed, but there were no factual affirmations of any type at the time for me to accept. When I myself saw affirmations, then I became inclined to believe this and other matters.

1928

Monism of the Universe

Prologue

Many of you will die during my years and I fear that you will leave this life with bitterness in your heart, not having learned from me—from an immaculate source of knowledge—that an uninterruptible joy awaits you. This is why I write this resume, while I have not yet ended these numerous preliminary tasks. I would like your life to possess a bright dream of the future, one that is a never-ending happiness.

What I preach, in my eyes, is not a dream, but a strict mathematical conclusion based on precise knowledge. I want to bring you into an ecstasy by contemplating the universe, by having an expectation of all these destinies, by the miraculous history of the past and future of each atom. This will increase your health, prolong life and provide strength to endure the vicissitudes of life. You will die with joy and having the conviction that happiness and perfection and infinity awaits you, and a subjective continuity of a wealthy organic life.

My conclusions are more comforting than the promises of the most optimistic of religions. Not one positivist can be more vigilant than I am. Even Espinoza, in comparison with myself, was a mystic. If my wine causes inebriety, then this nonetheless is natural. In order to understand me, you must completely sever yourself from everything vague, all in the areas of occultism, spiritism, debatable philosophy, from all self-proclaimed authorities, except for the authority of precise science, that is, mathematics, geometry, mechanics, physics, chemistry, biology and their branches.

Panpsychism, Meaning that Everything Has the Capacity to Feel

I am a pure materialist. I do not recognize anything except matter. In physics, chemistry and biology, I see solely mechanics. Know that the entire cosmos is solely infinite and a complex mechanism. Its complexity is so immense that it needs to restrains itself in some arbitrary, unexpected and accidental manner to provide the illusion that conscious entities possess a free will. Although we note that what occurs is periodical, but nothing actually is ever strictly repeated.

I call sensitivity the capability of the organism to sense pleasure and discomfort. We will note that when utilizing this word people often ascribe to it responsiveness or instant reflexes. But responsiveness is completely something else. Responsiveness is the entire body of the cosmos. Just as all bodies change in volume, form, color, strength, transparency, and all other qualities dependent on temperature, pressure, light, and in general the effect of other bodies. Dead bodies are even often more responsive than living. So are the thermometer, barometer, hydroscope, and other scientific instruments, considerably more responsive than a human.

Every particle of the universe is responsive. We think that it is also sensitive. Let's explain this.

Of the animals known to us, the human is the most sensitive. The balance of known animals are less sensitive the lower the organism is on the development scale. The least sensitive are plants. This is an uninterruptible ladder. It does not end even at the border of living matter, because no such border exists. It is artificial, just like all borders.

The sensitivity of the higher animals we can identify as joy and sorrow, hardship and bliss, pleasure and discomfort. The perceptive sense of the lower animals is not this strong. We do not know how to define them and cannot depict them. The intensity of their sensitivity is close to zero. I am speaking on the basis that with death, or the transfer of the organic state into the inorganic, sensitivity terminates. If it terminates due to a person fainting, or the heart suddenly stopping, then how much more does it disappear with the complete disintegration of the living organism.

Sensitivity vanishes, but responsiveness remains even in a dead body, except it remains less intensive, and accessible more for a researcher, than for a regular person.

A person can describe his joys and pains. We believe him, because he has sensitivity just as we do. The higher animals with their cry and movement force us to aver that their feelings are similar to ours. But the lower animals are unable to do this. They only flee from whatever poses a threat to their safety—tropism. Plant life for the most part is unable to do this. Does this mean that they cannot sense anything? The inorganic world is also powerless to communicate anything about itself, but this does not mean that it does not possess a lower form of sensitivity.

Only the degree of sensitivity of various particles of the universe differ and continuously change from zero to an undefined large magnitude, and this occurs in the higher entities, that is, those more developed than the human. They either derive this from people or else are located on other planets.

All is continuous and all is a whole. Its composition is the same and likewise its responsiveness and sensitivity. The degrees of sensitivity depends on the manner matter is combined. Just as the living world, due to its complexity and perfection, presents an uninterrupted ladder, descending to dead matter, so is the strength of sensitivity presented as the same type of ladder, not vanishing even at the borderline of life. If responsiveness, the display of mechanics, does not terminate, then why should sensitivity terminate, the display that is incorrectly called psychism. (We ascribe corporeality to this word—psychism.) And this and other displays proceed parallel, agreeably and will never abandon either what is alive or what is dead. Although, from another side, the quantity of sensations from a dead organism is so minute, that we conditionally or approximately can count it to be absent. If a white speck of dust lands on a black sheet of paper, then this is not sufficient justification to call it white. The white speck is the equivalent of the sensitivity of a dead organism.

In a mathematical sense the entire universe is alive, but the strength of sensitivity is displayed in all of its brilliance only among the higher animals. Each atom of matter senses the surrounding conditions. Reaching a highly organic entity, it lives their life and senses pleasure and discomfort; reaching the inorganic world, it as though sleeps, falling into a deep unconsciousness, into oblivion. Even if located in one living body, wandering throughout the body, it lives the life of the brain, then the life of a bone, a hair, a nail, skin, and etc. This means that it thinks: it lives as an atom confined in a rock, water or atmosphere. Then it sleeps, not cognizant of time, then it lives momentarily as a lower entity, then it recollects the past and draws a picture of the future. The higher the organic entity, then this perception of the past and future proceeds further.

I am not only a materialist, but also a panpsychist, recognizing the sensitivity of the entire universe. This quality I count indivisible from matter. All is alive, but only conditionally do we count it alive, meaning, only that which has the capacity to sufficient feel. Just as all matter always exists, under benevolent conditions it can pass into an organic state, and then we can conditionally state that inorganic material in its embryonic state is potentially alive.

Three Bases of Discernment

For our deliberations we possess three principles or three elements: time, space and energy. All the rest evolves from them, even sensitivity. These three understandings are indigenous only to the higher intellect and result from it, that is, the construction of the brain.

The simplest to understand is time. It has two directions: past and future and a defined magnitude, that is, it is measurable and it is also infinite, that is, not having a beginning or an end. I want to say that in the universe, there is as much time as a person wants. Every atom is generously provided with time. All immense periods of time—the known and imagined—are a total zero in comparison with its reserve in nature. The greatest gift of the cosmos for every particle in it is unending time, and this applies likewise to the human.

Space is a more complex display. It does not only have several directions, but still we ascribe to it various forms, volumes, and etc. In nature, space has no limits. It is just as abundant as is time. This means that an atom is additionally gifted with an inexhaustible and unlimited amount of space.

Even more complex is the display of energy. It evolves from an understanding of time and space.

These three elements of discernment are abstract, that is, they do not exist in the universe separately, but all of them blend into the materialization of matter. They define it. Without matter, not time, not space, not energy, can exist. Matter defines the three by using these definitions. They, of course, are totally subjective. (We consider it of little use to investigate their substance.)

The Law of Frequency

Some in general reject infinity. But know that it needs to be one of the two: finite or infinite. There cannot be a middle opinion. To allow the limitation of either in any magnitude is also not permitted. This means that there only exists one choice, and that is infinity.

Initially people thought that Earth was all that existed. All the balance consisted of heaven, and which had nothing in common with Earth. The stars, moon and sun were all deities. Then science explained that there are over a thousand planets that huddle just around our sun, all of them similar to Earth.

Likewise they thought the sun was the only one. It, for the most part, was promoted as the principal deity. But in addition to the sun, they found several thousand million of them, and in no way less in quality than ours. But since they are all surrounded by hundreds of planets, just as our sun is, then the numbers of earths increased to a hundred billion. This group of suns with their entourages of planets is called the Milky Way and composes a spiral nebula, and this is how we identify it. From a sufficient distance, this conglomeration of suns that compose our nebula are, in reality, seen as a hardly noticeable speck, should we be able to view it in this manner.

At the present time, about a million of such nebulas can be counted.

It is obvious the result is that the number of planets in the cosmos still increases a million times and now reaches to a quantity of 10^{17}— almost a million trillions. Our imagination and intellect will suggest to us that the discovery of a million spiral nebulas or Milky Ways composes only one group, one astronomical unit. I have the privilege to call it the *Ethereal Island*. Know that this quantity only occupies an infinitesimal part of the entire spatial expanse. So it the balance of the infinite portion just empty? Where spatial expanse exists, there must also exist matter. But since spatial expanse is unlimited, then the expansion of matter is likewise unbounded.

We come to the conclusion that there is no end to these ethereal islands. Their group composes an amount of the 5th exponential: one hundred thousand. The quantity in the series of astronomical units is likewise believably infinite, as is time and space, that is, existing in the 6th and 7th exponentials: one million and ten million—meaning, without end.

Frequent Displays of the Cosmos.

In General, the Universe always Possessed One Appearance.

People are inclined to think that all dies, just as they die themselves. They mean death as the eternal extinguishment of life or composition. This is one of the illusions of the mind, called anthropomorphism, or the ability to identify an incident in a person's life with his surroundings. The anthropomorphist thinks that some kind of stick, mountain, blade of grass, or insect, thinks and acts just as he does. For example, a rock is born, grows and dies; the mountain thinks, bacteria imagines, an amoeba connives, and etc.

But there is no reason not to believe in the return process—restoration. Is not the birth of plants, animals and the human a process that is a reversal to dying? Rather we only see, or want to see, the disintegration of the organism. However, creation as a reverse process reigns equally. It is even stronger than dying, and because the number of organisms on Earth is continuous growing. The population of Earth can increase with the preservation of the greatest prosperity, and a thousand times. If it were not for the limitation of solar energy that is absorbed by Earth and entering its confines, then its entirety could be transformed into a totally living mass. The entire planet to its very nether regions could then come alive.

After such a portrayal, is it possible to doubt in the vitality of matter? The brain and soul are mortal. They dissolve at their end. But atoms or their parts are immortal and so matter can be again rearranged and restored and again provided life according to the law of progress, and in a more perfected form.

Planets were earlier shining suns and then extinguished. This is what our present sun awaits. They must extinguish. The emanation of their rays—the source of life—will terminate, and all life on the planets will die. The universe will become like a dungeon without windows or doors. But is this possible? Will this continue forever? The universe has lived through an infinity of time already, and if suns did already extinguish, then there would not be thousands of billions of suns, those which we right now see in the many Milky Ways.

Rarely do astronomers observe the reduction of suns, for the most part they concentrate on new suns appearing. Every century several of them appear in the Milky Way. This is proper and such also pertains to other spiral nebulas. Bursts of newly-generated suns also appear there. This is the resolution as to the eternal radiance of the Milky Way and millions of other such spiral nebulas. Although suns extinguish, new ones ignite in their place.

So the sun has been shining billions of years, enlivening the matter on cold planets. They will eventually cool and cease emitting rays. But the atomic reaction within them does not stop for a long while: they accumulate radioactive material that converts into elementary and low density materials. All of this ends with an explosion, that is, with a display of a temporal star and the formation of a dispersed mass of gas and which in the course of a trillion years evolves into another sun with its planets and satellites.

Spiral nebulas also die, awesomely distant groups of suns or Milky Ways, and eventually their suns will merge together into matter of a very low density.

The merger of stars of each Milky Way is inevitable, and based on the theory of probability, it requires an immense amount of time, more than can be counted. It is a billion times longer than the interval of life of just one solar system. After the merger and compaction, a period of radiance occurs, and then following it is a cooling, and further will be an explosion, the formation of a nebula, that is, an embryonic Milky Way. But it resurrects or again provides a Milky Way composed of groups of suns. The hundreds of thousands of spiral nebulas serve as proof of such a restoration, those that do not exit from the field of sight of gigantic telescopes. If some of them extinguish, then others will be generated from unseen extinguished ones. The same with groups of Milky Ways. The ethereal island must have a temporary end. But still there does exist a large number of ethereal islands. If one of them transforms into elementary materials, then another from similar materials will materialize. All astronomical units live and die in order to again materialize, or more correctly stated, they only rearrange, while the forming is complex, they are more elementary objects and provide us the appearance of a star-filled sky, or else as low density gases that are hardly visible.

Each [orbital] period of a solar system consists of a numeral of the 3rd exponential (one thousand years), the [orbital] period of the Milky Way is astronomical, a trillion years (10^{12}); its lifespan is prolonged for a quadrillion years (10^{24}), while the lifespan of an ethereal island is sextillion (10^{36}). The more complex the astronomical unit, the higher its exponential degree, and so the period of its frequency is longer. So what is the result? The conclusion is that the universe, in general, has always presented one and same picture. Although our planetary system, a million years ago was in the form of a spiral, but the appearance of the Milky Way, over the course of a quadrillion years, has remained the same. This was the accumulation of hundreds of millions of suns in various stages of growth—from planetary spirals to cooled (on the surface) and dark suns. Although a quadrillion years ago the Milky Way was a body of extremely low dense material, but other Milky Ways existed in the ethereal island, consisting of conglomerations of suns, and its picture, on the average, has almost not changed over the prolonged course of sextillion years. Likewise the ethereal island was temporarily destroyed, but the group of them, an amount in the area of a hundred-thousand, lived as it did before, consisting of a large quantity of surviving ethereal islands. Each one maintained a million Milky Ways. Each from the previous in their order consisting of a hundred million solar systems, while each solar system consists of hundred of planets.

So the universe always contained a large quantity of planets, each illuminated by the rays of its respective sun.

The limited human mind cannot encompass the entire infinity of the cosmos. But let us imagine what we are able to observe in just our ethereal island over the course of a sextillion years. What will we then see? In every Milky Way, of which ours is comprised, suns have extinguished many time, spirals are forms and transformed into gigantic suns and so into planetary systems. And our solar system must have many times died and reappeared.

So many more trillions of years pass and we see how suns gradually merge into some Milky Way. They draw close to forming a single entity and after many trillions of years traversing a preliminary period of unimaginable brilliance from the collision and merging of suns they then draw toward extinguishment. So we see, after a quadrillion years, the extinguishment of Milky Ways, their reversion into spirals and a new structurization into the form of a gathering of solar systems.

The ethereal island is maintained for a sextillion years, meanwhile Milky Ways many times have dissolved and reappeared. But finally these Milky

Ways merge. Eventually the ethereal island is dissolved in order for a new one to appear in all of the brilliance of its life.

So what will we consider to be the beginning of the universe? If the ethereal island was to be limited, then we can accept structurization of the island in the form of amorphous matter as its beginning. But we must not forget that this beginning is only the beginning of a period and will be repeated an infinite number of times. If the Milky Way is to be considered the cosmos, then the beginning of the universe will be its generation from gases. And this beginning, of course, is only a beginning, one of infinitely repeating periods.

Finally, if the world limits the solar system, then the beginning of the cosmos will be its structurization in the form of some very low density material.

The Periodic Structurization of Atoms and their Placement among the Spatial Particles

The universe is composed of simple and complex atomic molecules. There are about 90 familiar simple atomic molecules, although in reality there are more. There are also billions of complex atomic molecules, and personally, their actual number is unimaginable. They are composed of the simple atomic molecules. But what we perceive to be simple can still be transformed into something even more simple, that is, having less than an atomic weight. At present in science there is enough basis to quantify 90 familiar simple molecules, all of them derived from hydrogen. Astronomy confirms the same. Embryonic suns, that is, planetary spirals, contain very few of the most simple elements. Then they develop into suns containing simple atomic molecules, those known and unknown to us. So what a person on Earth saw with the greatest of effort only during the past century, nature has been repeating uninterruptedly from the beginning of time, although slowly.

In reality, from the more simple or single-atom matter the entirety of diverse form is developed, those which are called chemical elements and their atomic unions. And in reverse, with the blast of extinguished suns and the formation of planetary spirals, simple atomic molecules are produced from the disintegration of the complex. Nonetheless, both processes always proceed simultaneously, but one will dominate at one time in one place, and the other at another time and place. Dissolution (analysis) dominates with complex items in suns, while the unification (synthesis) of simple matter occurs at their inauguration.

While I am describing the transformation of astronomical units, all matter is not only being changed or altered, but simple atomic molecules are being transformed into the complex and in reverse. I want to say that gold, lead and other elements are transformed into hydrogen and helium, and in reverse: hydrogen, helium and other simple atomic molecules, those having a low atomic weight, into gold, silver, iron, aluminum, and etc. Likewise I want to express that the central parts of the heavenly molecules fall on their surfaces and the reverse. In short, everything continuously and periodically alters and transforms. This process of volume and transformation of elements is always in operation, and without regard to catastrophic events. All suns release and lose matter, and then subsequently they will receive it. The brightest of them will lose more than they receive, while the dark ones operate in reverse. The planets are not removed from this process, of course, and which always possess even the smallest degree of radioactivity.

Monism

We preach monism in the universe, and no more than this. The entire scientific process consists in this drive toward monism, to a wholeness, to an elementary principle. Its success is defined gradually with its attainment of wholeness. Scientific monism provides the conditions for the structurization of the cosmos. Did not Darwin and Lamarck strive for monism in biology? Is not this what geologists also want? Physics and chemistry draw us toward this direction. Astronomy and astrophysics prove the unity of the formation of heavenly bodies, the similarity of land and sky, the uniformity of their objects and the energy inherent in rays emitted. Even the science of history strives toward monism.

In biology the cells of lower organisms unite, creating animals (the brain being its soul) with one purpose: to unite people into a society, a merger into one potent body. Soon the entire Earth must also unite in the same manner. The highest result will be such a unification on other planets.

I will add the potential capacity of each atom to live, even under complex conditions, within the known forms of unity and the general sensitivity of matter. The brain thinks, but atoms—those that compose it—feel. The brain dissolves and the ability for atoms to feel also disappears, and this is replaced by the sensation of non-existence, one that is close to zero.

It is impossible to reject the original appearance of organic life on such developed planets as the Earth.

There is much that is similar between the planets of the various solar systems: they are composed of one and the same substances, those of sufficient magnitude have seas and atmospheres, are illuminated by rays of their sun, are subject to the law of gravity, possess time intervals such as days and years.

So why did life not generate on them as it generated on Earth? What is true is that those planets distant from the radiance of the sun are cold, while those close to the sun are hot. But every sun possesses many planets. Some of them need to be located at a specific distance from their light-source, as is the Earth, in order to be conducive to life. Then theories indicate that all the planets originally separated from the sun, initially they were close to it, and only then gradually distanced themselves. So now any planet after some interval of time, possessing the conditions of proper temperature, is conducive to the generation and development of life. Likewise, every planet at some time, and including Earth among them, did not possess such conditions. Also, every planet, possessing right now an appropriate level of warmth, in time will lose it as its distance from its light-source increases. Other than this, the central light-source, whether igniting or extinguishing, provides all the planets appropriate intervals for the development of life, independent from superficial changes or its distance from the sun.

Some planets are small and so have no liquids or atmosphere. Others are large and so have not yet cooled. But when they cool, they are at such a far distance from their sun that they no longer receive a sufficient amount if its rays. This is why life is not conducive to planets that are either large or small.

Among other things, we note that such intervals can be prolonged for billions of years, which is sufficient for generation and development of organisms. Life, conceived at an opportune time, does not perish even with changes in conditions, since these changes occur gradually and life is capable of adapting to them.

There is no atmosphere on the smaller planets and this interferes with the formation of conditions to sustain life. We will not debate the issue right now. The basic conclusion is the following: the majority of mature planets or those planets with a gaseous atmosphere either are or were or will be inhabitable.

What Can We Expect from Humanity?

It is difficult for us to imagine how the process of life development occurs on some other planet, not being accessible to Earth. However, what can we expect from the population of our own Earth?

The human has traversed a great route from dead matter to a one-cell organism, and from there to its present half-way living state. Will it now stop at this point along the route? If it needs to stop, then not right now, because we see how giant the steps that science, technology, improvements and social structure are taking in their progress at the present time. This directs us to notice changes occurring within them. In any case, these changes need to occur.

Meanwhile, it is true that the person has changed little himself. These remnants are animal passions, instincts, mental weakness and routine. As far as social development is concerned, it even yields to ants and bees. But in general, it has advanced ahead of animals and subsequently, has powerfully progressed. Nothing will immediately stop and the human will not stop in his development the more that his intellect will motivate in him the drive for moral perfection. But for the meanwhile, the animal inclinations are stronger and the mind has not yet been able to dominate it.

We can await a soon introduction of an intelligent and moderate social structure on Earth, which will be compatible to its characteristics and its limitations. Unification will be introduced, the results of war will then be terminated, since there will be no one against whom to wage war. A happy social arrangement, guided by geniuses, will compel technology and science to proceed forward with unimaginable speed and with such speed in order to improve humanity's existence. This will subsequently cause an increase in growth. The population will grow to a thousand times its present, as a result of which the human will become the true master of Earth. He will transform deserts, change the composition of the atmosphere, and widely exploit the oceans. The climate will change based on what is needed or desired. All the world will become inhabitable and produce large harvests.

Complete freedom will be allotted for the development of the social, as well as the individual, qualities of a person, and nothing to harm others will be permitted. It is difficult to imagine the picture of the emotional peace of the future human, his security, comfort, understanding of the universe, tranquility and joy and assurance in a cloudless and unending happiness. Not one billionaire right now can possess anything similar.

Technology of the future will provide the possibility to overcome Earth's gravity and travel throughout the solar system. They will visit and study all of its planets. The imperfect worlds will be abandoned and their native populations will move elsewhere. The sun will be surrounded by artificial residences, their material being acquired from asteroids, planets and their satellites. This will provide the possibility to maintain a population up to 2 billion times larger than the present population of Earth. Some of their excessive quantities of people will be transferred to these heavenly colonies, while some of the initial migrants will increase their own population. This increase in population will occur at such an awesome speed that the manufacture of spermatozoa and ovum will be required.

Surrounding the sun, in an orbit that is near the asteroids, billions of billions of entities will grow and be perfected. They will receive a nature that is indicative of perfection: adaptable to life in various atmospheres, under various gravitational fields, on various planets, adaptable for survival in desolate regions or in low density gas, and the ability to live either with food or without it. In short, survivability solely on solar rays. They will be creatures that can tolerate heat and cold; and tolerate extreme and significant changes in temperature. Among other traits, they will be a superior and perfected type of organism, capable of living in the ether and sustained on direct solar energy, like a plant.

The settlement of other solar systems of our Milky Way will begin after the settlement our solar system. It will be difficult for a person to leave Earth due to its gravitational pull. It will be easier to overcome the sun's gravitational pull in light of the freedom of movement in the ether and the immensity of the radiated energy of the entire sun, and which a person will be able to utilize. Earth will become the point of departure for the settlement of perfected persons in the Milky Way. If they should discover some region on a planet that is desolate or underdeveloped or distorted, they will painlessly liquidate it and replace it with their own new world. If good harvests can be expected somewhere, they will remain there for cultivation.

The settlement of Earth has endured a difficult route, a prolonged road filled with suffering, but still much time for laborious development remains. Such a route is undesirable. But Earth, located in its particular spiral nebula—that is, the Milky Way—removes this difficult route for others and replaces it with an easy, exceptional effort and one that will not take a billion years, which is required otherwise for generation.

The Settlement of the Universe

What we have the right to expect from our planet, with the same right we can await the expect.

The conception of life appeared at the opportune time on all planets with an atmosphere. But on some of them, having the advantage of maturity and conditions, it flourished more splendidly and faster, providing the entities there technical and mental potential to become the source of a higher life for the other planets of the universe. They became the centers of promulgation of a perfected life. Such streams met each other, not hindering each other, and populated the Milky Way. All of them had one goal: to settle the universe with a perfected world for their general welfare. Is it possible to disagree over this? Along this route they encounter embryonic cultures, as well as those distorted and lagging, and those developing normally. In some places they liquidated life and in some places they allowed it to remain for development and general renovation. In the vast majority of situations they incurred a lagging society in the form of shapeless creatures, worms, one-cell organisms, or something even lower.

There was no reason of any kind to await a billion years of torturous development from it and the production of conscious and intellectual entities. the capability of multiplying the more perfected species—considerably sooner, easier and painlessly—is already prepared. I think that we on Earth will not expend time waiting for a human to evolve from a wolf or some bacteria, but we would be better off to increase its best representatives. So did the sowers of the higher life deliberate. In some places they destroyed the rudiments of a primitive life, while in other places they awaited a good harvest and renewal of the life of the cosmos.

They did this without bothering to settle other Milky Ways—groups of suns or spiral nebulas. This occurred in all the ethereal islands and in the entire infinite universe. In this way it expanded to the furthest regions of neighboring suns and quickly supplemented infinite desolate regions, and without having to go through the process of self-generation.

Intellect and the higher social organisms have reigned, reign now, and will reign in the universe. Intellect is what leads to the eternal prosperity of each atom. Intellect is the highest or the truest self-consciousness.

Worlds in the cosmos that are in an infantile stage, such as Earth, compose a rare exception. Are there very many people on Earth who only live a day? So there are few such worlds in a infantile stage. In the universe there are especially a small number in view of that the majority of settlements are performed through the means of immigration. A prepared, perfected form of humanity has already settled the cosmos. So what are the conclusions? We see an infinite universe with an endless number—decillions—of perfected worlds that have painlessly received a population and settlement.

Such centers of life, like the Earth, compose an extremely rare exception—like children at one third of their growth. A suffering life like that found on Earth is rare, as it has generated on its own and not through settlement. The concept of settlements dominates the cosmos as the preferred process. It can be compared to a person raising carrots and apples from seed each time, or how difficult it would be if a person had to continually start with autogenesis every time. This is possible, except a person would have to wait millions of years to acquire a carrot.

But in the cosmos, autogenesis is allowed, even though it is extremely rare, but it occurs as an inevitable means of renewal or to supplement that which has been perfected.

The role of Earth and a few similar planets, although of the suffering class, is honorable. Earth's perfecting stream of life is predestined to supplement the loss of regressive races of the cosmos. A weighty assignment has fallen upon Earth's population: this is its destiny, a supreme task. Few planets receive this, hardly even one in a trillion. But it cannot be otherwise. The opposite situation would contradict the intellect of the perfected, that is, the highest self-consciousness. But they are still able to cause harm to themselves.

A certain amount of suffering in the cosmos is a bitter inevitability, in light of the possibility of a reverse development of entities—regression or a backward route. But we can state that the organic life of the universe is found in a brilliant state. All that lives is happy, but this happiness is difficult for even a human to understand. The picture of Earth's disorders is a eyesore that is often suppressed, and perhaps you will state, "How can this be, if there is so much misfortune on Earth?"

So should we call the snow—black, on the basis that on its surface a few microscopically small black specks of dirt are found? And sometimes snow will reflect so much light that it is even painful to the eyes. Do we fear losing at gambling, if we win a quadrillion times and one ticket is blank? We do not even fear dying this year, even though death will affect about 2 to 3% of the population.

The Atom and Its Parts Have Sensations

We have seen that each atom, that is, every part that composes the substance of the cosmos, during times of upheavals, in the course of endlessly repeating periods of large and small intervals, in all the astronomical units, during the course of their dissolution and restoration—is reorganized. The atom decays and its weight or mass decreases, and it then recreates, as a result of which its atomic weight increases. The heavy or massive elements are transformed into lighter, and the opposite. This is an indispensable condition for the periodic cycle of suns, Milky Ways and other astronomical units. One is tied with the other. If a rearrangement of elements did not exist, then a course of periodical recycle of astronomical units would not exist.

Other than this, the atom migrates. For example, from central parts of some heavenly body it advances to the edge, from the suns it migrates to planets and then returns, and all of this occurs repeatedly without end. Based on this, it is obvious that there is not one atom that had not accepted innumerable opportunities of participation in the life of higher animals. So, entering the atmosphere or the soil of some planet, in time it enters into the composition of the brain of higher entities. Then it lives their life and feels the joy of a conscious and cloudless existence.

But we need to allow the demand of a billion years of incarnation for such an event. Nonetheless the atom participates in life an innumerable set of times. Since time is endless, and no matter what colossal intervals we may impose, the number of these phases is limitlessly great. To say this in other words, the life of an atom's participation in the higher life—and in general there is no other—in an absolute sense, is endless. This participation will never terminate.

But you will say that in a relative sense the atom is small, such that short periods of life are separated by billions of years of non-existence. But non-existence is not taken into the account of subjective time, as though it does not exist. It exists something like a higher consciousness and a never-ending happy life. In order to better understand this, I will speak on absolute and subjective time and its progression.

Absolute time is what every living—meaning not dead and not sleeping—organism equally observes. Absolute time is defined by clocks, the rotation of heavenly bodies and other natural or artificial chronometers.

Subjective time is something else entirely. This is time that the organism seems to experience. With one and the same organism it can move at various speeds. It can move faster or slower dependent on surrounding impressions or affects, psychological state or capacity of the mind. It depends on the character of the brain's activity. During sleep it moves faster. But here its movement also depends on the amount of dreams experienced during sleep. In deep sleep, without dreams, subjective time is almost unnoticeable. When a person faints, time is also unnoticeable. But often and with an abundance of happy impressions, time passes unnoticed. We do not seem to note how fast happy hours pass.

We will not speak of subjective time of various animals. Among the higher animals, its course is possibly the same as it is with a human, but this is difficult to ascertain with the lower. It is possible for their life to be similar to a person's dream. They have a weak understanding of time. They possess no past, no future, but it is only the moment that exists for them. The subjective time of the lower animals is not very interesting to us.

Likewise uninteresting is the concept of the subjective time of the dead, that is, the atom in an inorganic body, the atom in the atmosphere, water, field. This is all the more dominating form of matter. This is more than unconsciousness, so the course of time here is completely almost unnoticeable. Not only years, but billions of billions of years fly like a second in a living entity. Knowing this, should we really fear time in the inorganic realm or even consider it?

No matter how short are the periods of life, repeated an infinite number of times, they in sum compose an infinite amount of subjective and absolute time. It will never terminate. The past is likewise unlimited, as is the future.

This is why the course of an atom involved in some higher life is not just unlimited in regard to time, but is subjectively continuous. It is continuous also in the sense that the feeling of life will never terminate, although with each death or birth a sharp change is experienced. An atom is always alive and always happy, disregarding the absoluteness of the immense intervals of non-existence or its state in non-organic matter. For the dead, time does not exist and it cannot be considered. But since incarnation is inevitable—in the light of all that is said and in the light of endless time—then all of these incarnations subjectively blend into one subjective-continuous beautiful and unending life.

So if this is applied to a person, animal and each atom, what is the conclusion, and what is the finished condition in which they are found?

With the destruction of the organism, the atoms of a person, his brain or other parts of his body (likewise the eventual exit of the atom from the organism, which occurs many times still while he is alive), decline initially into an inorganic condition. Calculations indicate that approximately a hundred million years are needed for it to again materialize as a human body. This time passes for it as though it was zero. It is not subjective. But the settlement of the Earth during such an interval of time totally transforms. The earthly sphere will be then covered with only the highest forms of life, and only they will utilize our atoms.

This means that death terminates all suffering and subjectively provides an immediate happiness. If time incidentally passes considerably longer—trillions of trillions of years—then this is no worse for the atom. The Earth is no longer, the atom incarnates on another planet or other residences of life that are no less as beautiful.

The probability is infinitesimal that an atom on Earth will reincarnate within a few hundred years and so migrate into the composition of an animal that has not yet died or some undeveloped human. Life during the growth of the lower entities is not taken into account, as though it is almost imperceptible. Life in the higher organisms is similar to sleep, while life in the higher animals, although even terrifying on occasion to face—from the point of view of the human—is subjectively unconscious. A cow, sheep, horse or monkey, do not feel its humiliation, although the present day human feels his humiliation. As a result, the higher entities look upon the human with pity, just as we do at a dog or rat.

Such a rare accommodation of the existence of an atom in a contemporary human serves as a motivation towards his self-perfection and the liquidation of all other remaining species of lower life.

A Portrayal of the Tangible Life of the Atom

It is difficult to clearly present the tangible experiences of the atom to the person who has not contemplated his entire life on what was above proposed. But we will propose a comparison in order to understand and value the uninterruptible and infinite life of the atom.

We will ignore only the infantile period of the development of organic matter, which term and significance is unnoticeable in the cosmos, just as a microscopic speck of dirt is unnoticeable on a mirror or on a snow-white sheet of paper.

Imagine that our entire life consists of a series of joy-filled dreams. The person awakes, thinks for a second about his beautiful dream and hurries to sleep again, in order to delve back into bliss. In each dream he forgets who he is, and in each dream he is a new person. In one he imagines himself as Ivan, in another he is Vasily, then someone else. The second dream is not a continuation of the first, and the third is not a continuation of the second. Likewise each dream is independent of the others.

But the joyful facial expression indicates that all the dreams are beautiful. They reach only joy and never are to terminate. They always were, are and will be.

So what more do we need? You are continually happy! This is easy to imagine, a picture that provides an understanding of the life of each atom of the universe, wherever it may be located. This is just an elementary portrayal I suggest of the tangible fate of all and each. When you speak of this to people, they seem to be dissatisfied. They definitely want the second life to be a prolongation of the previous. They want to see their relatives, friends; they want what they previously have lived through. "But I will never see my wife, son, mother, father," they sorrowfully claim; "then is it best not to live at all." In short, your theory provides me no comfort.

But how can you see your friends when they are just a figure of your imagination created by your brain, and which will soon be necessarily destroyed. Not a dog, not an elephant, not a fly, will see their relatives for this very reason. The human is no exception. The one that dies bids farewell for all time to his environment. It is all in his brain and it eventually decays. The brain will again materialize when atoms again arranged in another brain. It will provide the environment, but another one and not having a connection with the previous.

Of course you are happy with your enchanting dreams, awakening each time with joy, in order to delve back into them. What more do you want? Right now you want to meet with the dead, but death will destroy all these wants. You dissatisfaction is only during life: one life is gone, so is the other.

We had friends in the previous life, but will we see them in the next life? This is what we will say, "Supposedly I knew him very well, although I never saw him and heard nothing of him." The new life will retain only happiness and contentment.

It is so difficult to sever ourselves from routine and grasp truth. It is likewise difficult to sense the movement of Earth's rotation.

There is a difference between a series of beautiful dreams that are not tied to one another and creating a single unit, and a series of active lives of an atom. These lives are conscious, as the life of sages, not vague or purposeless as are dreams. During the interval between dreams we feel something, although not for very long: particularly, it is life.

We do not feel anything during the intervals between the lives of an atom, disregarding their immensity. They do not exist for us. The life of an atom is subjective, uninterruptible, beginningless and endless, such that all of its many individual lives merge into one. Each of its individual lives is imagined in the form of a wave in an endless series of waves. There is not torment after death.

The active life—the wave—has a beginning and an end, and this is one period of a number of them. All of these periods, in essence, are sufficiently uniform: happiness, contentment, cognizance of the universe, cognizance of your never-ending destiny, knowledge of truth—which is the right route to a support of the cosmos in its brilliant state of perfection.

But every wave has a beginning and end. This is generation and extinguishment, conception and dissolution. This is repeated in the subsequent wave. However in no manner should you consider these waves identical. The waves can be very long, that is, the period of one life can be very great—prolonged for hundreds and thousands of years, but for the atom it does not matter, it is all the same, since the short waves of life merge together and emerge as infinity. Only the life of an organism should not be short, so that the entity, during this course, could attain a high development and produce abundant fruits. We should not shove something into it, as into an expensive machine, since this will break it and shatter it into pieces and within just a couple of hours.

Conclusions

Any planet of the universe with its moons composes an astronomical unit of the first rank. For example, our Earth or Jupiter or the others. Every solar system of the cosmos, for example, our planetary systems, is an astronomical unit of the 2nd rank (this is not to be confused with the astronomical measure, which is something else).[1]

Any spiral nebula, consisting of hundreds of millions of solar systems, composes a unit of the 3rd rank. Such is our spiral nebula, that is, the Milky Way.

We will name the ethereal island, consisting of hundreds of thousands of spiral nebulas, an astronomical unit of the 4th rank. There is only one ethereal

[1] The A.U., which is the distance between the Earth and sun.

island known to us, but according to the law of the repeatability of units, there should be many, and the conglomeration of them composes a unit of the 5th rank. According to this law, various ranks need to exist.

Proceeding further, we will briefly repeat what now seems to us to be verified:

We cannot deny the wholeness or some variations in the structure and formation of the universe: the wholeness of matter, light, gravity, life, and etc.

We cannot deny the general continuity of the universe, because in place of extinguished suns, new ones will appear.

We cannot deny the number of planets being infinite, because time and space are endless; where these exist, so does matter.

We cannot deny that a portion of the planets are located under conditions that are conducive for the development of life. Their number is infinite, because any part of infinity is also infinite.

We cannot deny that on some planets, animal life has attained a higher development that is superior to humanity, that it has advanced beyond the development of life on the rest of the planets.

We cannot deny that this higher organic life attains a great scientific and technological potential that permits the population to expand not only within its own solar system, but into neighboring ones, the remaining ones. (I have compositions about airplanes, dirigibles, rockets, about the wealth of the universe, about solar energy, and other topics.)

We cannot deny that higher life spreads and expands in the immense majority of incidences by the route of propagation and settlement, and not by the route of generation, as we have on Earth, because this delivers them from the delays and sufferings of a gradual development, because the intellect of conscious entities understands the advantage of the capacity of the cosmos to settle the cosmos. So the new regions will not be settled by wolves or monkeys evolving into humans, but by the increase of population of humanity itself. We will acquire fruits and vegetables not by evolution of bacteria, but from existing and perfected plants.

As a result we cannot deny that the universe is filled with higher conscious and perfected life.

We cannot deny that the atom at one time is simple and at another time is complex, that it periodically accepts the form of all the chemical elements.

We cannot deny that astronomical units are periodical, for example, the sun cools, then dissolves, returning to a low density mass that again will ignite and become a sun with planets. It will repeat the same further on. During this course, its matter is reorganized, while the chemical elements merge into one another.

We cannot deny that an atom is essentially capable of sensing life when it enters the capacity of an animal's brain. In this manner, it must subsequently live the life of various creatures.

We cannot deny that the tangibility of an atom does not vanish even in inorganic matter, although it is close to zero and such a state can be called non-existence.

We cannot deny that during the action of the rearrangement of objects and the transformation of chemical elements, there is no atom that does not periodically accept a participation in organic life, that is, does not end up even in rare occasions over the interval of trillions of years in the brain of higher entities.

We cannot deny that time for the atom in an inorganic object almost does not exist; that time in such a state for it is similar to non-existence, similar to unconsciousness. Subjectively, there is no means of deciphering the length of time since it is sometimes so immense.

We cannot deny that subjectively all comparatively short moments of the life of an atom in the brain of a creature merge into one continuous life.

We cannot deny that there is no benefit for the existence of an atom in the cosmos in any imperfect creature, similar to our monkey, cow, wolf, deer, rabbit, rat, and such. And likewise there is no advantage of its existence in imperfect persons or other creatures similar to them in the universe.

We cannot deny that all intellectual creatures will attain to the cognizance of this concept, not allowing any imperfection in the cosmos.

We cannot deny that the perfect is stronger than the imperfect, and as a result it is motivated by true egoism to liquidate painlessly all that is imperfect and suffering. Self-generation will be permitted very rarely for the restoration and supplementation of a regressive higher life. Such perhaps is the suffering and honorable role of Earth.

We cannot deny that the painful termination of life of the imperfect species is advantageous to the atom.

We cannot deny that, as the result of this, the atom can migrate only into a higher entity. There is no other route. Subsequently, its infinite existence can only be cloudless, intellectual, conscious and happy.

We do not notice a gray speck of dust on a snow-white field. In the same manner we do not notice the large number of planets that are doomed to the torture of evolution.

We will repeat this in other words:

Can we really doubt that the innumerable quantities of planets are periodically illuminated by suns?

Can we really doubt that on any other planet life attains a potential and perfection that for us—humans—is difficult to imagine? It permits them to overcome gravitational weight and so travel throughout the universe.

The subsequent result will be the expansion of perfection and the reign of intellect throughout the entire cosmos.

Can we really doubt that this was long since accomplished, infinitely far back in the past, and is the regular natural state of the universe? Generation or evolution with its sufferings is only a rare exception.

Can we really doubt that any atom, migrating into the brain, lives? It does not impose anything, does not dominate, but only senses life.

Can we really doubt that matter is continuously rearranging itself, periodically transforming every atom innumerable times, even if after immense intervals of time, and accepting participation in life?

Can we really doubt that the time that the atom resides in inorganic matter passes like a sound sleep or unconsciousness, and so time does not exist for it?

Can we really doubt that periods of life subjectively merge into one whole, as if being a continuous, conscious happy and endless life?

So only truth, perfection, potency and contentment exist in the cosmos, leaving behind just a little for the balance, and hardly enough to count, like a black speck of dust on a white sheet of paper.

All of my numerous efforts—those finished and those unfinished, the published and unpublished—are directed toward one goal: to prove the short concepts here proposed, or to make a final conclusion—that in general the cosmos contains only joy, contentment, perfection and truth. The contradistinctive amount in the universe, due to its minuteness, is unnoticeable, but we are blinded by the proximity of Earth.

Does God Exist?

(June 1931)

And you and many others ask me if God exists.

Before answering this question, I need to know how you interpret this word. As many as there are nations, as many as there are persons, so are there different interpretations of God. Which one of them is compatible with scientific truth, and can there be another? Right now among the French, the days of the week are named after five planets and the moon and sun, which were at one time recognized as deities.

Such deities, it seems, do exist because we cannot reject the existence of the planets and sun, but the understanding of them in ancient times was inaccurate. If you bear in mind the Hebrew and various Christian denominations (Orthodox, Catholic, Protestant, and others), then their teaching presents only vague indications of truth mixed with a number of harmful errors, foolish tales and superstitions.

The ancient sages, the creators of religions, of course, based their conclusions on their speculation of nature. But their comprehension of the cosmos was very weak, since science—the codified accumulation of the knowledge of previous eras and times—almost did not exist at the time. It is understandable that their deductions, their ideologies, could not be complete or accurate. The disciples of these sages distorted the ancient teachings even more.

Could there have been, for example, perceptions of the evolution of the universe that were accurate, when even Earth was perceived by people as being a flat and undefined mass, when Earth was the center of the universe, while all else existed subservient to Earth?

The absence of physiology forced them to suppose the existence of a special feature—a soul—in a person, which animated him. When it departed from the body, then the body became dead or lifeless. Some later thinkers supposed that such a feature resided also in animals, while others felt otherwise. Even the very ingenious Descartes considered animals not having a soul, that is, they were insensitive robots.

The dilemma of the existence and appearance of the human motivated them to accept the hypothesis of the existence of an omnipotent entity who promptly created human from clay, like a sculptor, and supplemented him with spirit.

We need to create a scientific definition of God if we want to disperse this passage.

Is there not something else that we can arrange, upon which we depend, which created us, which provided us intellect and knowledge of the universe, which graciously relates toward its creation, and provides it eternity and happiness? If we conditionally call this—God, then we will find an answer to our question.

We arrange the universe. Its released gases formed the sun, and planets separated from the sun, and life was generated on the planets, which, developing and perfecting over time, created the human and the creatures above him. We ourselves, our thoughts, our actions, are the creation of the universe. Likewise all of our infinite past, and so the future, materializes its will. Our will is only the display of the will of the cosmos.

So we see that the universe with its millions of billions of suns and even larger number of planets, along with their inhabitants, approaches just in part to our definition of God. We even have the right to attribute to our conditional God such qualities that the universe possesses according to present scientific data.

All of it consists of an unchangeable number of eternal atoms of hydrogen, and this is conditional since hydrogen is also a molecule. The sun and planets and people and the more mature creatures of other planets are all derived from them. Based on this, if a part of the universe—the organism— feels, then the entire universe is capable of doing so. Now we can deduce that our God—the cosmos—is eternal, immutable, alive, and we are parts of it, and this signifies that we are similar to it. Can we cease to live if the whole is always living?

Death is only the migration of one state of matter to another, from one feeling to a potential of another type. This is a new grouping of hydrogen atoms. The simple formulates into a complex, and then the complex back into the simple. And this repeats innumerable times, such that time has no end, just as it has no beginning. Our God is the cosmos, as we are its parts and always have been and will be.

The general view of the universe is always the same. If one sun extinguishes, then another ignites. If some planets annihilate, then others form. If some intelligent entities died, then others are born. Based on this it is obvious that our God—the cosmos—possesses as a whole a constant structure. Some parts of the universe possess an intensive lifeform—entities, while others have a very weak lifeform, one not even noticed, without memory, consciousness or mind.

But just as matter—stuff created from atoms and hydrogen—uninterruptedly reorganizes through the explosion of suns and planets, and the creation of new ones, and etc, then there is not one atom of hydrogen that has not accepted participation an infinite number of times in some organic life. All of these lives as if blend into one endless life, such that the immense interval of residency in some inorganic matter—conditional non-existence—is not embedded in memory and as though did not exist. Subsequently, every atom of our God, and ourselves along with him—since we are His parts—as though live an uninterrupted intensive organic life. This is another affirmation that God and us, not only vegetate occasionally in life, but also live subjectively an uninterrupted organic life.

So what type of life of God is this, us being his parts as animals and plants? We know that on Earth animals and people suffer, cruelly kill each other, endure pain, die, suffer destitution, are subjected to catastrophes, and etc. Is this really the general course of life of the universe, or is there another?

Millions of billions of planets have existed long into the past and so their entities attained maturity, and which we will attain over the course of millions of years of the life that awaits us on Earth. This maturity is displayed in perfected intellect, a profound comprehension of nature, and a technical potential, which provides the residents of the cosmos the ability to access other heavenly bodies.

Intellect speaks to these entities, "There does not need to be imperfection and suffering in the universe, otherwise we will suffer ourselves and subject ourselves to this imperfection and damage. And so mature entities on all the planets, impress truth, knowledge, joy and strength everywhere they can, due to their technological potential. They do not permit matter to endure agony for hundreds of thousands of years, but provide a gradual attainment of maturity in the form of higher entities. Evolution is rejected as a prolonged road of pain, and is replaced by the increase of prepared and perfected organisms and their proliferation on other planets."

So now we have this deduction: God—the cosmos—does not consist of pain and imperfection, but the best for Himself and so toward us who are part of Him. Truly He—the cosmos—can be called the Father and He approaches our definition of God. Such a God definitely exists, since it is impossible to reject the existence of the universe, its sovereignty, goodness and perfection.

Mind and Passion

When I refer to the term "sensation," I define it here as a level of pleasure or displeasure resulting from some kind of impact, independent of its form. Sensation in this meaning can be divided into pleasant, unpleasant and indifferent. There are no others. They can also be called positive, negative, and neutral.

Examples of positive sensations are: a good feeling about oneself; joy due to some reason; the gratification of thirst, hunger or some other desire or passion. Examples of unpleasant sensations are: a bad feeling about oneself; sorrow, melancholy, pain, and etc. When we cannot call a sensation either pleasant or unpleasant, then it is labeled as indifferent, as though it is uncomplicated. A significant part of our life traverses sensations that are indifferent, close to being neutral. Rising in the morning, but not in our later years, we experience vigilance, the desire to live, which can cause a positive sensation. During the course of the day we become tired and the pleasant sensation—toward evening or earlier—gradually passes into an unpleasant sensation, although to a weak degree. Obviously there is an interval between one and the other that is accompanied by an indifferent or neutral sensation, when we do not really know whether something is good or bad for us.

Sensations can have different magnitudes. In reality, no matter the size, it will still be positive, negative or neutral. It possesses various strengths, from a very little impact to a very large, or theoretically infinite, although in life such does not actually occur, because every joy and suffering is limited. But the strength of a sensation is something that we are not able to measure. In time we will learn. In earlier ages we could not measure area, volume, force, time, light as power, quantity of heat, electricity, and etc. Eventually we learned to do this. And even then the majority of people still cannot measure magnitude.

The highest level of positive sensation carries the appellation of bliss. The highest level of negative sensation is torment or agony. The more complex the entity, the greater the amplitude and span of the impact of sensation. Pleasant as well as unpleasant sensations can possess immense force. The superior creatures of other worlds that we can imagine perhaps possess a greater amplitude of sensations than humans. On the contrary, the lower creatures of Earth possess lesser powers of both positive and negative sensations. At the border of the organic world, the amplitude is close to nil.

Sensations can be diverse in form yet identical in magnitude. For example, the pleasure that comes from looking at a pretty picture or landscape can be equated with the pleasure derived from some kind of sound, like music or singing. In another respect these sensations are not comparable. In geometry, different forms can have the same area or the same volume, and are called isometric. A square and circle appear different, but their areas can be the same. The figures of a wolf and monkey are incompatible, but their volumes can be the same. So negative sensations that are dissimilar can still have the same impact. For example, pain from a burn can be equal to the pain from a laceration. Yet the impact of psychological suffering can always be greater than the impact of physical torment.

Just as adding two numbers that are the same except that one is positive and the other is negative will result in zero, so a step forward and a step backward causes you to remain in the same spot, and just as well a positive sensation when united with a negative result of the same scale produces a neutral state, that is, an indifferent sensation. Suppose I incur some minor sorrow. But then, suppose I need to satisfy some deprivation, for example hunger by eating a good meal, then the sensation from the sorrow is ameliorated and may even migrate into a positive sensation. When it migrates into an indifferent sensation, this means that both opposing forces of feelings are equal in strength. Based on this, the effect is called comfort.

Suppose that we experience a joyful sensation of a constant strength. Its quantity, obviously, is proportional to time and will have a continuous positive magnitude. With a constant painful—or in general an unpleasant—sensation, we likewise will be affected by a negative quantity of sensation. In practice, the strength of sensations change without interruption, at one time positive, at another time negative. The sensation of a living entity can be well expressed by a curve, when the abscissa (horizontal axis) is defined as time, while the ordinate (vertical axis) is defined by the strength of the sensation. The beginning of the coordinate is the passing moment, the right side of the curve from the vertical axis is the future; the left is the past. The curve above the horizontal axis pertains to a positive sensation; the curve below the horizontal axis pertains to the negative. (See the graph.)

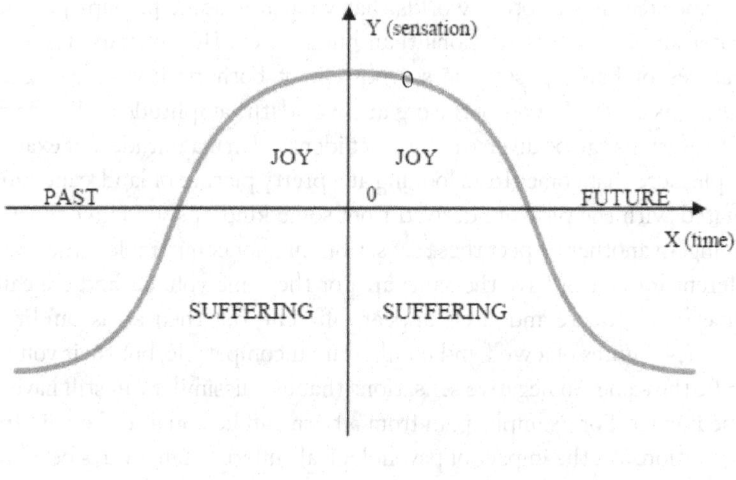

The magnitude (Co) of sensation depends on time (Bp) or is a function of time: Co = F(Bp). For a very small interval of time (dBp) the sensation can be considered constant, and so a small quantity of sensation in the course of this small interval (dBp) will be: F(Bp)dBp.

The quantity of sensation (Ko) from time (Bpl) to time (Bp2) can be expressed by the sum of the productions of these quantities during the entire interval, that is, the defined integral. Particularly:

$$Ko = \int_{B_1}^{B_2} F(Bp)dBp.$$

For example, from conception (B1=0) to death (B2 = Д = length of life), is expressed as:

$$Ko = \int_0^{Д} F(Bp)dBp.$$

Sensation changes over the course of the day, over the course of a year, and over the course of a lifetime. This is a general periodic (that is, iterative) sensation. They vary with children, adults and elderly. For the meanwhile I exclude the effects of passions, desires and various influences. For a child the morning is the most joyful part of the day. But the strength of sensation weakens and can shift into a negative, if something interferes with the child's sleep. Sleep is likewise pleasant, especially toward the morning when it is accompanied by pleasant dreams. The sum of sensations—the integral—of a child or adolescent to a certain age is more than just neutral or zero. For those who are older, the morning remains pleasant, but the burdens of daily responsibility reduce the sensation toward the evening.

Sleep begins initially as a negative sensation, but then shifts into a positive. The sum of sensations over the day decreases, but nevertheless it remains positive. With the addition of years of personal age it reaches zero. Beyond this point of neutrality it incurs a period of extinguishment, when the quantity of sensations is expressed with a negative number. At this time the morning begins with a negative sensation that increases toward evening and continues into the night. Sleep is accompanied by a heavy sensation or unpleasant dreams.

The reason for the relative morning vigilance and happiness consists in an increased activity of the brain as a result of a good night's rest. If we did not have nighttime, then the exertion of strength would be continuous, just as with the activity of the brain. If a person did not sleep, there would not exist a difference in sensation over the course of a sunny day, and this would just as well apply to people who spend time under artificial light. But nighttime and darkness forced the prehistoric person into inactivity, the result of which is sleep and an accumulation of strength. Any involuntary activity causes imbalance and inevitably results in an imbalance of sensation.

The annual rotation of seasons pertains more to the moderate regions. Spring and summer summon their warmth, abundance of experiences and food to strengthen the nervous system and physical activity and so is accompanied by an abundance of pleasant sensations.

Rotations in the course of time of a larger life should not be expected, because life does not repeat itself. It is as though we possess one wave: adolescence, old age, and then eternal rest. So what is this one wave? Childhood, youth, and maturity to a certain extent, as we all know, in general, is pleasant. Here the sum of sensations is positive and so provides life a certain value. Old age is not happiness, as we all also know. In general it provides a negative sum of sensations. Death is usually accompanied by torturous agony, which all the more increases the negative integral of the second half of life. The reasons for these displays is understandable. Adolescence is accompanied by an uninterrupted increase of cerebral activity as a result of the acceptance of new ideas, the contents of which accumulate up to a certain point during middle age. Following this is the weakening of the nervous system due to the weakening of the entire organism, the decrease of absorption and the inevitability of the gradual loss of cerebral faculties.

All that is stated above pertains not just to one person, but to the entirety of mortal entities on Earth or in outer space. In reality, who can be unfamiliar with the joy of young animals, the indifference of the mature, and gloom of the elderly?

The pinnacle of life's wave, the highest degree of joy, the maximum sum of the positive sensations of adolescence, and the negative of old age, suffering, the span and amplitude of life, depends on the ability of the brain to select one or another quantity of ideas. The greater this ability, so is life's amplitude more significant. But this ability in its own capacity depends on the arrangement of the brain and its magnitude. The arrangement of the brain we will leave off to the side as a little-known item and we will only turn to its magnitude.

I direct our attention to magnitude, because the volume defines the number of nerve bundles and subsequently the ability to select impressions that can be converted into ideas: memory, thought, imagination, and etc. The very lowest creatures of the organic world do not at all possess a clear nervous system, such as plants, bacteria, infusoria, and the like. In others we see 2, 3 or more ganglia, such as worms and insects. A 3rd group possesses a central nervous system with a large number of ganglia. Even more do we see among the animals who have brains in their heads, where the number of ganglia is in the thousands or even millions. The amplitude of life, the capacity of joy and suffering, obviously, is proportional to the complexity of the brain and its magnitude. At the edge of existence, meaning at the very conception of organic life, this amplitude is near to non-existence, almost as it exists in non-organic nature. Then it increases more and more. With the human it attains its highest degree, but it does not stop at this point. In other worlds, or as will occur on Earth in time, more complex entities can exist. There the amplitude of life is more significant: the waves are higher and longer.

Animals whose nervous system is not totally developed possess little joy, but at the same time their suffering is also little. At the lower level of the organic world one and the other are close to zero. Dead nature possesses almost an absolute zero.

This enters my mind: the sum of joys of every creature during the entire course of its life, it is not equal to the sum of sufferings during the same period? Then the full quantity of sensations or integral of all life, no matter how complex or simple it might be, is also equal to zero. In reality, A-A=0. If it is this way, then the conclusion is that life, that is, its sensation, is only an excited zero,[1] non-existence brought into agitation, a calm derived out of balance. It is possible that we are mistaken. However, why is there no happiness in old age and a pleasant death? The opposite attitude seldom occurs. A person is not always able to communicate his sensations and torments any more than an inarticulate animal.

A person can hardly believe in the possibility of a painless death. It is not without good reason that people who possess an imagination fear a painful death. In general no one can assure another of a joyful death when having to suffer any kind of execution. Death by an electric surge generates greater terror than hanging, decapitation, or poisoning by morphine or some other poison. I speak of the sensation of death, and not about the fear of non-existence.

The understanding of such deaths is unreliable and is only the result of fantasy and our ignorance. If the emergence of the brain and the idea of joy that is attached to it and which generates this joy, then its destruction—death—obviously must be accompanied by a quantity of suffering that is equal in magnitude to the amount of joy that evolved from the brain. Life presents an entire cycle: evolution from soil and its return there. Whatever was provided it will be removed.

Let this hypothesis be a risky one. I myself do not have the capacity to believe it, yet all of it has been derived from many preceding and useful conclusions pertaining to a beneficial life.

Calm and non-existence reign in inorganic nature, although in its mathematical—that is, a precise—sense, this cannot be: an atom is one of the stages of a simple life. Plants and the lower entities, like bacteria, infusoria, worms, insects, invertebrates, creatures composed of cartilage, are likewise near a sedentary state, and they hardly even sense their own demise. Their life is not only unconscious, but it passes as though a dream. The smaller vertebrates likewise have a meager sensation of life. But the greater the magnitude of their brain, the stronger their sensation.

The pain of death of such creatures, as the dog, horse, cow, the pig and its wild breeds, is many times weaker than the human, and relative to the magnitude of the brain, the dearth of their faculties. But yet they still deserve

[1] Or, an excitement of life resulting in zero

humanity's compassion. Sometime during the infinite life—since the atom wanders from brain to brain—the human will also inevitably experience the same as them. Such menial creatures can be compared to a well-built mechanical robot, whose capacity of sensation will not cause an increase in pain due to a poke of a needle or a cut in the skin.

Time will eventually make the human the master of Earth. He will arrange the life of plants and animals, even determine their individual destinies. He will transform not only Earth, but its living organisms, not excluding himself. Based on what was just stated, he can unceremoniously identify with the lower entities, destroying those harmful to him while increasing the population of the good. Our heart and conscience will feel safe. Such entities almost do not suffer anyway.

The species of unconscious entities with large brains must be mercifully annihilated by terminating the possibility of reproduction, since a direct destruction would be cruel. But it would be allowed for predators such as the wolf, tiger and such, if no other means are readily available.

But how to transform the human? What is he to be? Consciousness requires not only a special construction of the brain, but a significant magnitude. Such a brain is definitely accompanied by a large amplitude of sensations, as well as a significant pain in death, and which is not desired. A small brain is limited in its capacity and easily drives the entity into an unconscious state, because it provides little intellect and comprehension. Then it becomes a source of suffering for itself and others, as we see in the animal world and, alas, the human, which at the present could still use a good dose of conscience.

The perfection of the human brain is demanded, but without any decrease in consciousness. During this process, either an increase or decrease of the brain's magnitude will also occur. The latter is most possible since a large volume of the present human brain is occupied with unnecessary qualities and even those that are harmful to people, lusts for example. This might even be to the human's advantage, since it will also reduce pain during death and in general the amplitudes of life. Pain during death can be decreased to a desirable level, and at the same time increasing the length of the process of demise.

During every stage in our growth, and even over the course of a single day, we experience vigilance at one point, then some burden later, then a joy and then some sorrow. Joy, of course, is the desirable, but on what does sorrow, melancholy, bad moods, and burden of life depend? Can we not estrange then?

We saw that there exists probability to consider the general sum of sensations of every living creature during the course of one life's period as equal to zero. To say this in other words, there is as much joy as there is suffering. So it is obvious that our sorrows and the difficult sensations of life have a source in our joys or evolve from them. They are the reason for our sorrows. In reality, if we rejoice much in life, then we must in general suffer the same amount. If we have little joy, then our suffering is less. With the dead, since they have no joy, there is no suffering. But in part, this is not completely clear. In reality, life as though provides us with many joys for which we pay nothing: appetite, sex, and others. But the matter is that the satisfaction of every passion is preceded by the pain of desire, that is, a demanding sensation.

There are two categories of sources of demanding sensations: natural, those which a person at the present cannot escape; and unnatural or artificial, those which a person can escape by his own strength or by the intervention of another.

The initial source of our sufferings is our passions. To experience a joyful satisfaction of hunger or thirst, or sexual requirements, and many other desires, they need to be preceded by a prolonged depriving sensation, even if not very strong: weariness, exhaustion, boredom, and discontent. The organism is charged with sensations almost unnoticeably, but over a long interval in order to acquire much satisfaction in the end. Without such a charge, there is little joy in the satisfaction of the passion.

So to escape life's destitution, what is needed is particularly an unobstructed satisfaction of passions. Life becomes tranquil, content and happy. There are no greater joy than these, and no more difficult weariness. Such a life is desirable for the majority and must be provided it. This is the right of all workers and those incapable of working, both the maturely developed creatures and the weak.

In practice, almost no person can escape all the difficulties of life—even with a soon satisfaction of passions. In reality, we first need to define the majority of these passions that need to be uninterruptedly satisfied. But such a continuity is impossible. A person is so constructed and so lives in a manner that he gratifies these desires in spurts over a defined internal of time. Over their course there occurs a charge with a subsequent pleasant and quick discharge. In short, the interval between the short momentary quenching of passions is saturated by the difficulty of life, a fatigue of spirit.

During the stage of adolescence, these fatigues are tolerable. In old age or illness their impact is greater. With children, fatigue weakens with their capacity of acceptance and creation of new ideas. With old age, fatigue initiates the quenching of ideas or the gradual demise of the brain.

It is possible to explain the mechanism of this event, that is, the reason for a short joy and a preliminary prolonged fatigue.

Let's suppose that the body of some animal needs water. It is the brain that usually contains, and because of its complexity, the majority of desires, the majority of thoughts, that is, the stimuli generating activities of various muscles, to satisfy such needs. But if a human is a robot, how will it fill its stomach with water? He will just die without this capacity.

Other than this, fatigue due to thirst is accompanied by the impossibility for a person to concentrate his thoughts, inability to work, anxiety and a repulsion to every type of work that will not lead to satisfaction of this passion, thirst in our case. This is understandable, since good nourishment benefits the parts of the brain that lead to a fulfillment all developing wants. Based on this it is obvious that calm and regulated work is impossible without the regular and opportune satisfaction of passions. From another side, their unimpeded and complete satisfaction attains a balance with routine work, but does not promote special and genuine development of the brain.

Indeed, unsatisfied desires and various obstacles force those various parts of the brain to increase in work, to seek conclusions that it earlier could not fine. In general it is the parts of the brain that were earlier atrophied or undeveloped and rudimentary that are developed first. So does a person received his special construction. He receives a comprehension of what is hunger, cold, and various others deprivations. He seeks their reasons and then means of estrangement. He becomes a special and useful activist, and not just an ordinary worker. Not every deprivation leads a person to such a comprehension, but only a wealthy mental nature. But such people for the most part are well endowed materially, never in need of anything, and never have some undeveloped part of their brain remaining. In their case, they need to willingly create some deprivation for their brains to acquire qualities that are valuable for humanity. The destitute do receive it, but they are in no position to be of help to humanity due to their weaknesses and lack of authority. For the most part it is necessary for the powerful and secure to experience deprivation and a trend will evolve from this. But it all has its limits.

A carefree and secure life is self-destruction, and excessive toleration and torment leads to the same. The matter is that with an excessive restraint of satisfying passions, not only is the general vain activity of brain restrained, but also the effort of all the parts and organs of the body: circulation of blood, respiration, digestion, and etc. The organism openly destroys itself and often arrives at a sorrowful conclusion as a result of excessive restraint of passions. So young men and women die due to not satisfying their sexual needs, the requirements of love and happiness. They perish due to the abundant need of satisfaction. Not all will fall to the wayside, but all suffer and weaken.

But all have a quantity and intent. Somewhere in the middle is the truth or something better.

So, it is not the sensation of thirst that summons activity that is inclined to satisfying that desire, but the nervous mechanism of the creature. The sensation of joy and suffering is an auxiliary product of life and its mechanism. It is not feelings that motivate the body, but the neuronal apparatus.

If atoms did not possess rudimentary responsiveness, then life would progress just as in the past: it would have no meaning, it would be dead, we would be robots. However, the difference between an automatic-doll and an animal pertaining to sensation is only quantitative. But the simple automation of plants, lower creatures and objects of artists, have so low a capacity of sensation, that they are practically the same.

The mechanism of passions force a person and animal to do what needs to be done to preserve their life and perpetuate the species. Any other sort of mechanism is impossible, because otherwise they would destroy each other or else not leave a posterity. But with the gradual development of the animal kingdom, over the course of millions of years, the mechanism has become more and more complex, such that intellect and will were added to the above and a stronger brain. Intellect has grown to such a point that now the person can—and this is obvious—exist without the lower means of existence, that is, without animal passions or instincts.

Let us imagine such a person. Suppose his body is emaciated due to lack of water. He sees this, but does not experience any discomfort. The cognizance of a near death or degradation of the brain can motivate him to get some water and drink it, although he may not sense any satisfaction from this.

He does not have a female lover. He knows that due to this he will not produce any posterity. But he does not suffer, his organism does not weaken due to his asceticism, but on the contrary, it works much stronger. However the thought that the population of Earth and the universe will perish without procreation, that life will migrate into an unconscious lifeless state, forces him to think about some offspring. The female can also deliberate in the same manner. So we have a means of continuing the species without interjection of passions.

Fire burns the skin and other things can do damage to the body. But it is intellect that can warn and preserve the person from fire and other harm.

It is understandable that a person could exist without passions, as long as he has a high intellect. He could live, procreate, and be happy, without the inclusion of passions. There would be no difficulties of life, but at the same time there would be no intense joys, short momentary indulgences, or the satisfaction or desires.

The lower animals possess a weak intellect or do not possess one at all, yet passions are inevitable, even though they are weak in them and hardly deserve to be called passions. The simpler the mechanism, the more unsophisticated the automation. And right now many people could not survive without passions, because their intellect and will are weak. But in time, a route for entities without passion, but with a high intellect, may just be produced.

So what would be the advantages for this to occur? These: an equilibrium of mood, absence of the difficulties of life, and more productive work. Other than this, passions that are not overcome often serve as a reason for very stupid conduct. Intellect will escape this, establishing a higher quality humanity.

The subject entity pertaining to feeling will possess only two cycles: a cycle of adolescence and development, when the number of ideas and activity of the brain grow, and a cycle of old age, when one and the other gradually vanquish. The first cycle will be accompanied by a tranquil joy that we can quickly identify as vigilance or productiveness. The second is a cycle of tranquil sorrow, but not terminating any capacity of work, but only weakening it. The gradual and prolonged quenching of the brain will deliver the entity from pain during death. As a result, life will extend for such a long interval and the corresponding negative sensations will be almost unnoticeable, and so humanity demandingly yearns for this destiny, this transformation. Other worlds long ago evolved entities with such qualities. The universe is full of them. Earth is the exception, because its maturity is still in an infantile stage.

The beauty of nature, plants and animals—as long as we can see and hear—submerges us into a world of impressions, serves as the source of education and new ideas during adolescence, or the increase of its vanishing, and all is pleasant. If we do not see or hear this beauty, we deprive ourselves of it entirely or partially and so experience a negative sensation. Under normal conditions of life, this difficulty is not a danger, since the person can always estrange it, especially living in a warm climate or arranging nature according to our desire: light, warmth, atmosphere, and others. What also excites pleasure in us are the movements and sounds of people, and especially those of the opposite gender. When this is absent, a person suffers. This is another source of suffering, but it can also be escaped with a good social environment.

There does exist a combination of movement and sounds—music—capable of exciting unusually strong sensations and which for the most part are joyful and even overwhelmingly pleasant. They can summon sorrow and tears, but these are tears of joy and pleasant emotion. Other types of music are dangerous, or alcohol or narcotics, and provide a meaningless bliss; they dissipate moral strength, and weaken the person for a long period. However, all of this pertains to artificial means of dealing with life's difficulties.

There is healthy music, which is the norm, which does satisfy the accumulated and unsatisfied passion for sounds. We feel an ecstasy when we hear the voices and singing of men and women. But can all hear them? Instrumental music is more accessible, although it would be better if it was exchanged by natural attractive sounds. Music in general relieves passions, not having any other release. But often due to the impact of sorrowful conditions some passions do not have an exit.

Other natural stimulators of the nerves and their subsequent deactivation accompanied by a loss of strength and difficulty of life are narcotics. Such are: alcohol (ethanol), morphine, cocaine, theine, caffeine, arsenic, and many others. Essentially these are poison-medicines and effective means of artificial stimulation, which in small quantities are useful only in a few situations of life under the conditions of some illnesses. Their use without good reason, meaning, for pleasure, serve as a means of subsequent anguish and pain, and also as a reason for the damage of your health, and premature aging and death.

We drink tea, coffee, and wine, but they are harmful and are the causes of colds, sickness, fevers, and other bad reactions. Smoking is likewise damaging; it is dirty, leaves a repulsive taste in your mouth, ruins the air in your apartment, as well as causes the brain to lose its capacity. Such items should not be touched or utilized. It is easy to restrain ourselves from them, as long as we ignore people who promote them. There is no need for us to walk along the edge of a cliff. A person who drinks wine and experiences enjoyment from one bottle of wine will want to repeat the experience. But then to attain the same experience a second time, he needs to drink two bottles of wine. The person falls further and deeper into a hole, until he crashes and mutilates himself.

Suppose I get drunk with wine. Everybody feels good initially, but the results will surface later, like a hangover or internal pain. Many persons who get drunk will feel valiant, courageous, self-confident, dissolute, bold, boastful, and in general develop a false sense of personal strength. How much evil can this attitude produce. What follows later is a decline in mood and the sensation of opposite characteristics which are accompanied by malice, profanity, violence, and various stupid actions. How much suffering and harm this causes to those surrounding you! And then after recovering from the effects of inebriation, after the temporal happiness, your need to pay for all of this with suffering, the inability to react properly and loss of health.

To be healed of such a popular passion can only occur through a route of intense effort, and there is no help from the majority who also seem to have the same problem. What this takes is inspiration, to develop a repulsion against wine, or to acquire medical help against addiction to morphine or cocaine and other narcotics. But this occurs gradually and over a prolonged period. To withdraw quickly will cause horrible side effects or force a person to suicide.

But there are also other causes of suffering that we have not yet discussed. This is illness, pain, degradation of the body, death, and loss of close associates. All know about this. The questions are the reasons for the causes, their meaning and whether it is possible to circumvent them.

Suppose they stab, lacerate, compress or burn our skin. We suffer pain and this compels us to estrange our self from the perpetrator. Essentially is it not the suffering, but a mechanism of the nervous system. Among the lower entities it is instinct; among the high entities is it the action of the brain. The lower the entity relative to the construction of the nervous system, the less the suffering.

An illness of the body or its traumatic damage—wound or accident or the like—requires complete calm and even inactivity for its recovery. Then its strengths can concentrate on one matter: increasing nourishment of the ill person or the damaged organ for its speedy healing. It is understandable that illness must be accompanied by weakness of all the organs, and any movement of the body can interrupt the recovery.

There is no point in estranging the reason for these sufferings. But the future entity molded by thousands of years of selective breeding, like the sugar beet, with a highly developed brain, will not suffer so much as do people at present. Indeed, just intellect will tell us that we should protect our body from harm and illness, and it alone can preserve us. It can also compel us—and without feeling any pain—to avoid perpetrators or to lie in bed and cease activity for some temporary period until we get well, stop the exertion of mental and muscular strength and provide calm to the sick organ, even if it should cause some boredom.

The loss of close associates or even their sufferings and failures, likewise cause us agitation. The reason is that our thoughts will concentrate on the suffering of those close to us while we forget about ourselves. The goal is to somehow assist or deliver our close associate. Subsequently, the brain in general withdraws some of its activity. In addition to this, if a close associate dies, an entire series of thoughts regarding him is extinguished and so we experience a negative sensation resulting in a weakening of the brain's functional ability. As long as he was alive, we were concerned over him, that is, we concentrated our thoughts on how to assist him. Here there was at least some kind of activity. When a person departs from the Earth's scene, we lose all hope, however we do not abandon our thoughts about him and this results in our inability to help him which causes depression and some personal suffering. Its strength depends on the degree of closeness of the dead to us, meaning, the amount of thoughts located in our brain that were concentrated on the close associate.

There are a multitude of other reasons for our joys and sufferings, but this is sufficient for the meanwhile. The reader can now deliberate on what he has read and add his own.

The Cause of the Cosmos

The following is not infallible information, but a mixture of precise science with philosophic deliberations. They can be accepted or not accepted. This is better than wandering in the pitch-black darkness of occultism and spiritism.

The cosmos is similar to a cinematographic screen, where a series of pictures is developed completely automatically. It is similar likewise to a configuration of sounds that provides us with a record player. It is similar to the future mechanism that will combine displays of light with sounds and other items, even displays of thought, as in an adding machine.

We know that there exists a cause for all these automatic effects. It consists in the human being a creator. He is himself something higher in comparison with his productions.

But can the same be likewise said about the cause of the universe, as we speak about the cause of artificial things? It is difficult to count the cause of the universe as identical with it itself. Indeed, a person in his creativity cannot create even one crumb of matter, not one drop of work, and due to the law of preservation of matter and energy. None of this is actually created, but the cause has provided it an entire infinity in the form of a limitless cosmos.

This means first that we can speak of a cause, that it is not only something higher than the universe, but also that it cannot have anything in common with matter. During various eras, during various stages of development and knowledge, people accepted various causes of this existence and the possibility of prosperity.

All that was useful or formidable was accepted at one time or another for the cause and it became the object of worship. Either beauty providing pleasure or a more perfect life, or a river irrigating a region with its canals, or awesome animals and imagined evil monsters that can destroy everything, or useful domestic animals and fantasy creatures that are good and upon which the contentment, satisfaction and health of people depended, or people who are heroes, or outer space aliens.

The more perspicacious thinkers understood that the sun provided the conditions for all of their prosperity and welfare. It provided grass, vegetables, fruit, bread. Without it the existence of domestic animals would be impossible. It provided a joyful light, warmth, and a multitudes of other benefits. The sun became the source of life, the cause for everything. It was the philanthropic cause.

But others, who were more intellectual, understood that light does not have a mind and if it could, it would be bad to the same indifferent extent as it would be good. Doesn't it also sear the harvest and dry it and cause sunstrokes. There was no sense to try to deal with an impersonal cause.

Other than this, astronomy opened an innumerable quantity of other suns, no less potent than ours, although not having the same significance for Earth, due to its distance. There was, obviously, a more important, cause, or a more serious source, for all the suns and all of our welfare.

First of all the error lies in that only part of the universe was accepted as the cause for all this providence, for example, the sun, moon, human, and etc, meanwhile it was so clear that the entire cosmos provided the conditions for our life. It is difficult to suppose that some part of it did not have an influence on us sooner or later. This comprehension led them to pantheism. The universe was accepted as the cause for all that is occurring. Further research regarding the cause for the cosmos are fruitless, so the pantheists think.

All depends on matter in its united entirety. It birthed innumerable suns, and even more innumerable planets and life on them. It produced the perfection of organic entities, estranged suffering and caused every atom to be happy. Now enough research on matter, as we are located in its hands. It is our mother and despot.

But now questions arise. Why is the universe geared to provide goodness and not badness? Why is it this way, and not the other? However, is it possible to imagine a different order, another construction, other laws of nature? Other than this, what use is it to venerate the universe? It is just an infinite complex mechanism. Veneration demands sacrifice. What use are such sacrifices to nature? There is no sense to make of this, either for it or for us. This equally applies to us, whether to venerate the sun or fire. What is needed are for sacrifices or our activities to be useful to us in their own right.

So for this cause the most wisest of people counted that which led all people and all other entities toward happiness, contentment, comprehension and eternity, as our sovereign and despot.

We speak of ideas, rules and laws, leading all that is alive—and of course all is alive—to a durable and endless satisfaction. Some accepted love toward humanity as such a fundamental idea; others accepted love and kindness to all that is alive on Earth; a third accepted kindness toward all earthly and heavenly, to each atom or its parts. Belief in this idea, in its redemptive abilities, its materialization in life, its promulgation, subjection to it—was not in any case a fruitless sacrifice. It could produce great fruits.

And so, the general higher love toward an atom, in the eyes of the wisest sages, was worthy of worship, that is, observance of the laws that expressed this love.

But know that kindness and laws evolve from people and other considerably higher entities. This means that we are arriving toward a veneration of select scholarly persons and other entities possessing the highest of qualities. To them is our obedience, attention and respect. Although these entities are offspring of the universe, but nonetheless veneration of the universe itself is fruitless.

It is useful for us, the limited and weak, to observe the will of the highest elect. It is very sufficient for us, it seems, to establish ourselves on this premise, and since here is the cause for the cosmos!

We can demonstrate that a purpose in allowing the existence of a cause, to ascertain its properties and have certain feeling toward it. A dog ran into the forest and cut its paw against a sharp fragment of a pot. It cried for a while and ran further along, doing what it wanted to do. Its brain did not consider the injury of major consequence and did not bother to turn any attention to the fragment.

Later a hungry and lost person was wandering in the same area. He sees the fragment, by which the dog was wounded, picks it up and closely examines it. The fragment of the pot leads him to developing the thought of a possible residence nearly. He was dying from exhaustion and so started to look for this residence. Soon he found it and was delivered from his dire situation.

So which one conducted himself in the more prudent manner—the animal or the human?

The archeologist, digging, gathering and studying the remnants of the previous life of organisms, makes conclusions about their existence, structure, development, properties, habits, tastes, form of life, and etc. So is the history of our predecessors compiled and the history of the evolution of animals. The stupid or barbaric person, due to his ignorance—and they are the majority—breaks some antique vase, or scatters unfinished instruments utilized by the ancestors—those that he accidentally finds—and disdains all of the half-decayed traces of their earlier existence. So we should not be like this animal or like limited and ignorant people, and we should ascertain the properties of the universe, to the amount that our minds and comprehension will allow us. This is because the universe exists. Perhaps someone will say to us, “If we seek the cause for the cosmos, then obviously, this cause will possess a new cause. So we will never reach the end." Of course, I will agree, but the mind is limited and so it will be good if we will acquire some knowledge about the primordial cause. There was a time when the topic of the cosmos possessing a cause was the boundaries of our deliberations. But even earlier it was limited by only the sun, or at other eras the Earth was the limit, or the objects on its surface. Progressing, we are taking steps forward, seeking the first cause for the universe.

Due to the survival of various implements and objects, based on them, we can make conclusions of the contemporary or past existence of humanity. Just as we base similar conclusions on the burrows, nests and bones of animals, so does the universe uninterruptedly cry to us regarding the existence of a cause. Just as we make conclusion of the properties, art and instruments of the sculptor based on the remnants of some antique vase, so can we ascertain the properties and goals of the cause by basing our conclusions on the cosmos and its properties.

But this cause cannot in any case be for purposes of evil, since the world in general is beautiful and provides its separate parts as almost an immaculate and subjectively-uninterrupted happiness. It likewise cannot be annihilated, since the world is unlimited and eternal, not having either a beginning or an end.

When we gaze at a well-designed and constructed statue or doll, then what enters our head is that the producer has capabilities beyond his product: the doll does not speak, not walk, not think. It cannot reproduce itself. It is obvious that the product stands below its producer. So by studying the universe, we must come to the conclusion that the cause immeasurably transcends the cosmos.

What is overwhelming for a person can be insignificant as a cause, like a toy for an artisan. What is beginningless and endless for a conscious living entity, can be a limiting factor in the cause, because on its own merit it is incomparable to its inclusive objects, like a pitiful pot or statue is incomparable with a person.

Perhaps someone will say, "What do the properties of the cause have to do with us, if we are totally dependent on the universe? It is enough just to study the universe."

But the matter is that investigating the properties of the cause provides unexpected and new deductions that just studying will not provide. All that is ascertained will have a beneficial effect on the actions of humanity and other conscious entities.

Qualities of the Cause

The cosmos is endless and beginningless in regard to time and space. This is striking. What is exceptionally striking is the manner that infinity was installed. Infinity is the product of a mind just as is the generation of the universe itself. This is something subjective. But what we would consider infinity can be a limited magnitude as far as the cause is concerned.

But we will never understand this. We can introduce only an example that will explain our idea, but it does not prove anything. A worm tunnels through an apple and does not see any end to it, nor any beginning, so it seems to it endless. So in the same manner does the cosmos seem endless to us. The endlessness of time and space is the act of a higher creativity. Just as we fabricate some type of object, so did the cause create the endlessness of all generations of existence.

The universe is endless due to its essential substance. And here we can repeat the same deliberations that pertain to matter, that is, the prevalence of ether, suns, planets and other heavenly bodies.

The cosmos possesses an infinite capacity of work: potential energy. The indefinitely burning suns can serve as an example. Although they eventually extinguish, but they or others will ignite. It is so abundant that even in a limited piece of matter or ether, the available energy will never exhaust. This third infinity is the same generation of the human mind as all the others. To use this infinity as a cause is very insignificant, like a toy for a person. So what is the actual cause that produces all of these marvelous derivations for people! It is relatively potent as an artisan in comparison with some unnoticeable speck of dust that falls from his clothes. We must note that all of our comparisons are meager as quantitative relations, that is, the cause is immeasurable higher.

The universe contains nothing except atoms with its parts. Every minute these atoms are prepared to generate into life. There is no atom that would not periodically accept a participation in a higher living organism, entities such as the human and higher. Mathematically, if we were to accept even the smallest unnoticeable sensation as a quantity, then all atoms are always alive. So the entire cosmos, to the its extreme limits—which of course does not exist—is always alive in the absolute sense. It always feels. What a high level of life-sensing capacity is the cause. We riskily compare it with the life-sensing capacity of the higher descendants of the human in relation to the sensing capacity of grass or bacteria.

The parts of the cosmos, the atoms, live billions of years, but they still decompose. However, even the infinitesimal fractions—the products of their decomposition—are eternal. They only play like waves in the sea. How consistent is the cause and what a game it plays!

What always existed, the universe for example, could not have been created. But know that this deliberation is subjective, it is the product of the brain. The world is created, but the human mind cannot comprehend this. What is beginningless for us does have a beginning as far as the cause is concerned. So there is no need to try to find a beginning in a circle. A may fly only live for a day. If it possessed intellect, then the life of a human would seem to it to be beginningless and endless. We will repeat: the world was created. All cosmic infinities are only composite parts of the item which desirably would create a cause. So how potent it is if the universe is only one of the items of the cause.

We must attribute to it the power not only to create, but also to destroy. Likewise to do this once and many times over, an unlimited number of times. The cause must have the capability to liquidate and reassemble matter. It is apparent that a person's present limited observation does not allow him to notice that the cause intervenes in the activities of the universe or reconstructs it. Not creation, not the annihilation of matter, is noticeable. The cosmos develops mechanically, but the right to create and annihilate cannot be removed from the cause.

And so if the entire cosmos suddenly disappears, then this would only be the display of the higher will. If a million new universes were generated, and not having anything similar to one another, and not with our world, then this would not be anything marvelous. What would be marvelous is if a potter was to smash his cart full of pots and recreate of them an amount of pots ten times more. The comparison, of course, is pitiful, just as with all the others.

Deliberating the cosmos is bewildering. It is constructed such in order to provide for itself only happiness. So the wisdom of the cause, of its fabricated item—the universe, causes us to faint from dizziness.

We proved in the composition *Monism of the Universe* that the cosmos is governed by intellect, and specifically its own, and that in the general picture we see nothing except for its completed facets. The life that it generates is higher than the human intellect. The human and the animal are the exception—the indispensable nursery of regeneration. Such planets, as Earth, are so rare, that they are almost ignored, just as dust on a white sheet of paper is almost not noticed. So the universe in generation is not able to retain sorrow or stupidity. It fabricates its joy and perfection on its own, completely natural and inevitable, and so inevitable that every live organism attempts to alienate itself from pain and other sufferings. It is only on Earth for the time being that the lower animals and even the human do not maintain enough strength or ability to accomplish this, but in the cosmos it is sufficient. In time there will be a sufficient amount and even in the distant future posterities of humanity. In short, the living universe, on its own, is content and smart. And if this is the feeling of the world, then what a state of health is its cause!

Its goal is to provide immortal and never-interruptible welfare. It is immeasurable not only because it—in relation to time—does not possess either a beginning or end, but because it is infinite due to its expansion in space and even perhaps in strength, in order to provide growth for everything. The cause, creating its toy, provided it joy in lieu of torment. We know that it is great. How much goodness does the cause maintain if it has provided even one of its toys such an amount of happiness that is beyond the ability of the human mind to grasp.

August 1925

The Scientific Ethic

Prologue

I will not for a moment depart from the idea of singularity—monism—and materiality in this treatise. The word *intangible* is everywhere placed in quotation marks and signifies specifically tenuous or rarefied matter, but still organizable. This will be explained better in the further text.

Can something of fantasy be considered in this treatise? Categorically, nothing. I depart from the principle of infinite complex matter that, in its own turn, evolves from infinite time, that is, from what the universe was always, and so eternally, comprised. If I was to point at the character, forms, number and other creatures of other worlds, then this would be actual fantasy. But I do nothing of the kind and have not.

To suppose for example, the existence of organic life on other planets as not being a fantasy, but to communicate some kind of defined character to them is now a fable, because we know nothing about them.

We must not likewise decline from allowing the organization of matter that is more tenuous, that is distant from us in the area of many decillions of years.[1] This is likewise not a fantasy. We see that in all the corners of Earth, matter is organized in the form of plants and animals. So why can this not also be in other worlds with a tenuous form of matter? Here is another example that is similar. If life did appear on Earth, then why cannot it likewise appear in the same manner on any one of the trillions of other planets that have the same conditions as does Earth? The appearance of organic matter can be rejected to a certain per cent, like 10, 50 or even 90, but to say in all planets is unreasonable.

[1] A decillion is 10^{33} years.

The population of the universe is absolute, although not factual truth. To say that the universe is empty, deprived of life, on the basis that we do not see it, is a cruel deception.

Everything Is Alive

The population of any planet can grow by means of birth. Only the surface area of Earth and the energy it absorbs from the sun limits the magnitude of the animal population. If the excess of the new-born could be relocated to other planets and find sustenance there, then an immense portion of the planets would be converted into life-sustaining regions.

Some of these entities are not included in the category of Earth's animals. So either these entities can be transformed into others that are capable of being living bodies on Earth, or the animals themselves can be transformed and then utilize the materials on Earth. Then the entire planet as a whole would be composed only of animals and humans.

It also operates in the opposite direction. This living world, under unacceptable conditions, will die and be transformed into a dead planet.

What is not seen in this picture is that all is alive, and it is only temporarily found in non-existence, in the form of unorganized dead matter.

You will say, "In practice, we see that a living organism lives barely one moment, after which it merges into the soil, into inorganic matter, for a trillion years. At some time the opportunity to again live will again reach it."

But is it possible for you to imagine conditions under which the subject mass will live uninterruptedly?

Let's imagine a transparent, sturdy cocoon filled inside with oxygen, carbon dioxide and nitrogen, and also containing a small amount of moist soil, plant life and a few living, intellectual creatures of both genders. The plant life, consuming some of the soil and atmosphere, produces offspring. This feeds the animals and then the animals digest the plants and the waste is converted to soil, which again, with the help of the plants, provides food for the animals. And so without end. This is no miracle. This same sequence occurs on every planet that is capable of sustaining life. Only in the portrayed cocoon the mass of animals composes a noticeable portion of all the mass of our isolated little world, while on a real planet it is insignificantly small. We notice that our little world, in general, is immortal, just as earthly life is immortal.

We can also imagine the type of creature for which the inorganic world does not play any role, that is, this creature has no need of plant life or soil or atmosphere. It sufficiently survives on its own body and the sun's rays.

Let's imagine an object, covered by a transparent, flexible skin, but not allowing any matter to penetrate it. Chlorophyll is located under the skin in some places, just as with plants, capable of converting carbon dioxide and other waste products into oxygen and other nourishing substances, just as in plants. Animals will then feed on such objects along with oxygen. Nourishment is now continually provided and all that is required is the sun's rays to form food and oxygen.

If Earth's humanity is immortal and our little world in this transparent receptacle is immortal, then why can't one creature in its transparent cocoon be immortal? Nature or a person's reason in time will be able to attain this. I am assured that the mature worlds beyond Earth have long produced such creatures: immortal and living solely upon the sun's rays.

So what is the advantage? Every particle of the universe, that is, every conglomeration of matter, can accept a form of a living and even immortal creature.

Of what does the universe consist?

More than all else, we see suns and which seem—due to their immense distance from us—as sparkling spots (stars) and even the many of them blend into one almost radiant fog. These suns are in the astronomical quantity of a million billions. There are so many of them, that if you were to divide them among people, then each would receive about a millions suns.

Suns are immense. At their assigned time, lesser bodies separated from them, similar to Earth. These are the planets. Due to their small size, they externally cooled and permitted the generation of plants and animals on them. The planets generated even smaller bodies, similar to our moon. There are more of them than planets. And planets and moons are the most diverse of sizes. Some are a thousand times larger than Earth, while others are a thousand and million times smaller. Some of the smaller ones are the size of a speck of dust. The smaller the size of the heavenly body, the greater their number.

All the heavenly bodies pulled to one another, like magnets, and soon they blended into one group, if their original movement, of course, was not so far from one another. The movement of the heavenly bodies evolved from their mutual attraction as a result of some incomprehensible force called universal gravitation. Every material item is subject to it, although for items of a small mass, it is infinitesimal.

Time, Space, Mass and Sensation

The human mind is constructed so that it cannot exist, circumventing these 3 concepts: time, space and mass. These 3 concepts characterize the universe, that is, they define the substance of which the entire body of the universe is composed. It is impossible to imagine the world or any part of it without time and space.

We are capable of comprehending an interval of time having a defined magnitude, for example, a century, year, minute, second. But all time does not need to be comprehended as being limited. There is no need to express the entire sum of time in the universe as a number or quantity. Not the past, not the future, has a limit and both one and the other is infinite. So time is a magnitude that has no form, and its sum in the universe is infinite. Time only has two directions: past and future.

Space has not only many directions—for example, right and left—and magnitude—for example, volume—but also forms, for example, spheres, cones, humans, butterflies, lines, surface. This comprehension is much more complex than time. The spatial volume of the universe, just as with its time, is impossible to imagine as some limited, measurable, expressible quantity, because its sum in the universe is infinite. To think of it otherwise is unreasonable, as with the limitation of time.

The wealth of the universe consists in its never having to terminate or slow time, and never, not in any direction to reduce or cease expansion. Time and space are eternal; they will never disappear; they are non-disintegrating. The existence of one and the other without substance or matter with the ability to provide a living form, with its joy and sorrow, is unreasonable.

It is impossible for you to imagine substance without time and space. Know that every substance or body occupies a known volume—space—and exists within a defined interval. This means that time and expansion are as though an imminent or inherent appurtenance of substance, an indivisible quality.

On the other hand, it is possible to imagine time and space without substance? For example, is it possible for you to imagine an empty spatial expanse and time? Such a question is rhetorical. But we think that space itself and time are as if composed from substance. In reality, for example, the attraction of every material particle expands its rays into all directions and encompasses the entirety of the infinity of the universe. This means that space, even if it appears to be empty, is filled with substance and comprises substance. Essentially, substance, space and time are the supreme mysteries, but we do not consider them solved. We think that since time and space are unlimited, infinite on every side, then so is substance.

Research confirms this. Due to the extent of development of observational astronomy, the limits of substance expand more and more. Recently, millions of special worlds have been discovered, each of which contains a billions suns, along with their planets, moons and a quantity of smaller bodies. Factually, but to the extent that our imperfect vision permits us when utilizing such instruments and through a hazy atmosphere, matter is unlimited.

Nonetheless it still all seems to us that its boundaries are expanding, and its boundaries as we envision them are likewise endless, just as with time and the spaciousness of the universe. There is no reason for time and space to have an existence without matter, and just as the former are limitless, then the expansion of substance in the form of suns and planets must likewise be limitless.

There also exists a quality of matter that is indivisible from it, as is time and expansion. This is the ability of matter to sense sorrow and joy, or its sensation, and this is discussed in other chapters.

Substance

From the beginning people discovered an innumerable amount of diverse substances and thought that they had nothing common among themselves. Every substance was unique. Such were the various minerals, metals, fluids and gases.

Then they noticed that some substances evolved from others and in the opposite manner, some of these again returned to their original state. So by heating red mercury oxide powder there resulted mercury and oxygen, and in the opposite manner, red powder was formed from mercury and oxygen. In short, the formation of new compositions was noticed with a forced combination of others. This subjected them to thinking that perhaps diverse substances were composed from a few basic and even perhaps from one basic element. Accumulating facts and research over the course of time brought them to the following conclusions:

All substances are infinitely diverse, but all of them are composed only from about 90 molecules that are called—simple. Their combination in twos and threes and etc, although not occurring all the time, but only under certain conditions generate all the innumerable substances of the world.

From the beginning all of this pertained solely to Earth, and then it reached to all the heavenly bodies. Suns and planets emerged as being composed of the same substances as Earth. It seemed to be impossible to confirm this, since not one heavenly body was accessible to people. However the small heavenly bodies that collided with Earth, falling to it in the form of meteorites, appeared to have the same composition as did Earth's substances. This similarity between Earth and outer space seemed unbelievable. As a result many deliberated for a long while that perhaps meteorites did not have an interstellar evolution, but were thrown down by volcanoes onto the ground.

So how are we to discover the substance of the sun and other inaccessible heavenly bodies? Even 100 years ago we thought that this would be attainable for a human.

All know that molecules can be heated or cooled, meaning that the one substance can exist in various states. As a result, many of its qualities change in these different states. So, under a high degree of heat—temperature—all solid and fluid substances become gaseous or vaporous. In addition to this, at a high temperature all substances radiate light and it appears as though uniform. But if you were to direct this light through a prism, then it separates into a long dark strip consisting of small lines of various colors, and each substance has its own peculiar arrangement, different from all others.

In the opposite direction, at a low temperature, all the gaseous substances condense and then solidify.

What was also noticed was that every gas possessed an inalterable physiognomy, which can be used to identify that specific gas.

But the universe for the most part is composed of substances similar to the super-heated forms of gases from the sun's surface. All of these gases emit rays and they reach Earth. When we pass them through a prism we recognize a number of gaseous physiognomies in the form of a number of colored lines. To separate them is not so easy because there are so many of them. However when we compare these portraits with portraits of Earth's gases, we see a similarity, we see a blend of the physiognomies or spectrums of the gases on Earth that are known to us. The conclusion we make is that all the radiating heavenly bodies contain the same super-heated gases that are found on Earth.

Since planets separated from suns, the consequence is that they must be composed of the same substances as their respective sun.

So the uniform composition of the universe from 90 essential and basic molecules is confirmed by the following:

> The light of suns and rarefied masses of gas;
>
> Meteorites;
>
> Uniform formation of suns, planets, moons, and all other heavenly bodies from immense masses of rarefied gases—nebulas.

The conclusion is that there is no essential difference between Earth and heavenly bodies. One and the other are composed of one and the same materials.

But science did not just stop here. It was about 100 years ago that the wise Brug[1] suspected that all of the known 90 molecules consisted of hydrogen.

[1] No knowledge of who this is.

Now this is more and more confirmed. The matter is complicated only by hydrogen being not a single item, but that it is composed of two parts: protons and electrons. This means that all 90 compounds are composed of them.

We now almost arrive at the concept of the singularity of matter: all is derived from the element hydrogen. But other than hydrogen, we also have ether, a substance strikingly rarefied and resilient.

This now provides us the means to suppose that there does exist still other forms of matter, more simple, out of which hydrogen is composed, and without which an indivisible or indecomposable matter cannot be derived.

So in general, what type of matter is this? Simple or complex? A certain proposition does exist of an envisioned simple, meaning non-complex, compound that is composed of fast-moving individual and identical particles like atoms. But they are so small that they cannot collide in the manner that billiard balls collide against one another. But all of them are connected to each other by a special means of attraction that—due to the very small space between them—forces each atom to depart from its path in the manner of a comet as it enters our solar system due to the sun's gravitational attraction. What occurs is similar: getting near but not contacting. They are not actually able to touch each other: it is an attraction to be near, but repulsion when too near. The relative strength of attraction between them—atomic pull—is many decillion times greater than the attraction of the heavenly bodies—universal gravity.

The close evolution of several of such atoms, along with their defined and very rare combination, forces them to amalgamate under such a pull, yet without contact, in tight groups: in twos and threes. In general, just a few atoms. Of course, the opposite sequence is possible, that is, the dissolution of the complex group into more simple elements.

So more and more complex groups were all formed in the universe, which a person, due to his ignorance, considered them indivisible—simple—and so named them—atoms. But since the beginning of the universe is infinitely distant from our era, even though there was never such a thing as a beginning and the world always existed, the process of amalgamation continued infinitely, and this is why all the simple molecules of matter or atoms that are known to us must have an endless complex arrangement and unknown construction.

Even hydrogen is complex with its protons and electrons and ethereal particles. We are no longer speaking of the molecules of the 90 compounds: their complexity is beyond doubt. Science seems to want to deal with the complex, and especially those infinitely complex molecules.

Population of the Universe

We see that all suns emit one and the same light, that all of them with their planets are composed of the same substances, and that even these substances have one source, one primordial matter.

Gigantic suns, created from rarefied gases, their condensation or precipitation, resulted in the formation of more and even more complex matter. These suns reduced in volume and rotated fast and so caused rings and planets to separate from them. Initially they were hot, but then cooled from the surface due to their relatively small mass.

We see a complete uniformity of worlds in the form of millions of billions of suns together with their planets, moons and other heavenly bodies. We need to add to this uniformity of light and substance the uniformity of the gravitational force, meaning, that all bodies possess this attraction. The planets of medium size possessed oceans and a gaseous environment or atmosphere.

So the question arises: Why does life not generate on other cooled bodies of the universe, in the same manner as it generated on Earth?

It seems that it is cold on some planets distant from their sun, while it is hot on others that are close to their sun. Some of them, due to their large size have still not cooled and so life cannot generate on them. The smaller planets and their satellites—moons—have no atmosphere or water—oceans—and so remain desolate. The orbit of others was exceptionally eccentric, the planet's axis inclined in the direction of the orbit. There are also a few that are little adaptable to life as a result of extreme temperature changes on their surface.

But not all of these planets are inadaptable. From dozens of groups of planets and hundreds of smaller, those that accompany every sun, at least one planet in every solar system can possess opportune and favorable conditions for the appearance of life. If we were to allow such a premise, then millions of billions of planets have this capacity and are populated.

Initially, composite complex compounds appear on every favorable planet. They become simple compounds once adapted to the chemical environment and effect of the sun's rays, but bacteria does not yet exist there. Then the more complex forms of bacteria appear, similar to those were have on Earth. A microscopic vegetative world is formed and it divides into two directions: plant and animal. One and the other develop simultaneously. The size of these and other creatures increases and their construction becomes more complex. Plant and animal life follow the pattern similar to what is already on Earth.

But not all of the simple entities proceed on the same route. Some, those that mature and adapt to conditions, remain close to themselves; others increase in size and maturity only to a specific degree. So on every planet creatures are formed of all sizes and at all degrees of development.

Here is one trend that seems to be the dominant. An entity surfaced that is more intelligent and stronger than the others, kind of like the human. Initially he was close to animals and ruthlessly exploited them and killed them, and even did not spare his own species. The strongest of the intellectual group also dominated and tortured others. But then intellect further developed. The supreme human comprehended nature and its strengths, started to utilize them and had compassion toward those similar to himself. This compassion then extended to animals. Finally he comprehended the subjective uninterruptibleness and endlessness of the life of every piece of matter. He understood that when he was malicious to other creatures, he was being malicious to himself in the concept of the unlimited life of his own future. He realized that his personal welfare consisted in not having any suffering and not any disorder anyplace in the entire universe.

So on every planet that has existed for a sufficient number of years, what remains are the various perfected species that have attained this.

The moments of a planet's generation—its departure from a gigantic sun—are the most diverse, and likewise the growth of a planet is just as diverse. The vast majority of them have a respectable growth that causes them to attain the perfection of an intellectual life, that is, a completely mature population. There are few of these such as Earth and due to its recent generation and so its growth is small.

Would it be proper to count the age of the human population of Earth negligible, until it lives to a million years?

The time of its historical life is a total of only 10,000 years. It still a trillion years of development ahead of it. Ten thousand years hardly comprises one hundred-millionth[1] of its destiny in the future life of Earth and sun. Isn't this just a second relative to one human life? So what is the conclusion? The entire universe is full of the life of perfected creatures, and which the Earth is awaiting and so are a few other planets that have not reached mature growth.

We spoke that the maturity of a planet is the most diverse and likewise the conditions of life. These thousands of billions of planets cannot be identical with respect to favorable life-sustaining conditions. Some were generated earlier and life matured upon them earlier than on all the balance. Other than this, the conditions for life could have been better than on the balance. They attained maturity prior to other planets, expressing in themselves a supreme potential and a long, painless and happy life.

Their intellect disclosed to them that the route they traversed—from bacteria to maturity—was a difficult route, the path of millennia of

1 1/100,000,000

sufferings and insanities. The perfected were able to complete this route, but other planets are still waiting to do it. Why should they endure pain? Isn't it possible to deliver other planets from this course of torment for their creatures?

The technical potential of the initially-generated bodies permitted them to overcome the difficulties of their planet, harness their sun's energy, and become dominant in their planetary system. Here they could liquidate the embryonic life of planets and replace them with their own mature population.

Then their technology proceeded further. They started to complete journeys to other suns of the universe. They arrived at planets of other solar systems, where the pitiful life of bacteria and slugs was only starting. And there this life was annihilated and exchanged with their own population. On other visited planets they met creatures of a somewhat higher development, such as reptiles and amphibians, and they were painlessly annihilated and replaced by the perfected species. So did they proceed on the majority of the planets of the universe, not seeing any need to allow the self-imposed suffering of such gnawing creatures to continue over the course of millions of years.

However the destiny of the smaller portion of planets, perhaps only in the trillions, was left to themselves. Observing them, but not annihilating their life, they permitted them to develop on their own.

Not turning our attention to the nursery-level of intellect among those planets similar to Earth, consisting approximately of the destiny of only a trillion planets, we can state that the universe is full of intellectual, potent, and happy creatures. Their genius and potency populated the universe and delivered it from the torment of self-generation. These creatures are similar to perfected people who will evolve from the present humanity.

In time it will organize into a beautiful and happy community under the direction of the very supreme and most worthy of among the people of the future. When he should depart from the scene, then another just like him will take his place, the best of all humans.

Every planet attains happiness, complete community development and rule by the most supreme of the entire population. But this will not occur by a route of self-generation, as this is a torturous route, but by the route of settlement and breeding through perfected creatures from other planets.

The planets are governed by their supreme presidents, their own category of planetary deities. But each planet, while increasing, will export the excess of their population to the planetary regions surrounding their sun. There they will build special residences, even more beautiful and comfortable, than their originating planets. The regions near the sun will provide not only expansion, but also solar energy, which is a billions times greater than what reaches the planets.

So will a special and abundant population surrounding each sun surface. Here also a better-arranged community will organize under the direction of the best of the creatures of the entire settlement. Such a community arrangement, and particularly the best possible, will arise while under the leadership of the settlement's presidents.

An association of the suns located near each other will be organized and they will elect a ruler. It is not known whether there will be an end to such unions and associations, and probably not, because they are so necessary. The universe has a need of them to decide: where to transfer excess population, where to locate them, at what solar system, with what qualities, at what distance, whether they should they await an upheaval of the sun, or consider if this will this threaten the population. All of this needs to be known, and without leaders of the universe, without an expanded guidance, association and knowledge, all this will be impossible.

So the creatures similar to a perfected human that populate the cosmos comprise organized and beautiful associations under the direction of the president with his vast quantities of assistants. On one planet there exists presidents of various levels. Not counting their assistants we will see rulers over planets, solar systems, groups of suns, Milky Ways and so forth, probably without end.

The final ruler is possibly over the entire universe, all of its infinity. He will serve as our deity, in whose hands we will always reside. The qualities of this deity, as we notice, are benevolent, but he will still be elevated in our eyes when we penetrate further into the secrets of the cosmos.

Life Is Subjectively Uninterruptible, There Is No Death

We saw that life in general is complete and beautiful. It cannot be otherwise, since the developed intellect of mature creatures introversively—egoistically—will not permit this. If it did permit it, then those who possess such an intellect would themselves be suffering the entirety of their endless life.

Now here is a question with which to deal. Will death, or non-existence in inorganic and unorganized matter after the disintegration of the creature, be torturous or painful?

In a deep sleep, when life is still far from extinguishment, a living entity feels almost nothing. Time flies unnoticeably. Sometimes it seems that ten hours pass as one second. A creature is likewise insensitive or non-reactive in a coma, when the heart is almost at a cessation. It seems that time in such a state does not exist. So how much more does time hide and vanish without a trace when, not only the heart, but the entire organism, disintegrates.

Time is a subjective sensation and applies only to the living. For the dead, it categorically does not exist.

So an immense interval of non-existence, or the residence of matter in a unorganized—dead—form, as though does not exist. What does exist relative to time is a short interval of life. All of this blends into one infinite whole, since the large intervals outside of time cannot be counted as zero.

Of course, the one and same piece of matter incarnates, that is, accepts a state as a living entity an infinitely large number of times, since time never terminates.

But all of us mistakenly think that our existence continues only as long as our external form is preserved, as long as I am Ivan. However, after death, I may become Vasily, although this will no longer be me, but someone else. Myself, I vanish forever. What vanishes is just our superficial form. But life is reinstalled from the same particles of disintegration and sensation continues in another Vasily or Petro, or a lion or fly, or some plant.

Sensation depends not on form, but on matter. A bird flies, continually changing the location of its mass and sensation. So you are Ivan, but imagine that nature or art creates something else out of yourself, but it is not you. In this manner does life continue infinitely.

After death, the material of our tangible composition dissipates and so sensation is as though dissolved and erased, yet the material in its small parts of the former creature now becomes the new material of the bodies of other living entities.

Let's look at this indirectly. The life of every present-day person or animal is composed of parts of matter that lived at some time in the past in the most diverse of places. However, this did not interfere with it appearing as a new life. So although the material of our composition will be dispersed after our death, this will not interfere with it again living.

Now let's deliberate on this in a more substantial manner. Here is a living entity. So what parts of his body contain the quality of sensation? We see a large number of animals of the most diverse size and mass, and each one of them senses. Based on this it is obvious that this ability does not depend on the magnitude of the animal. So every organized mass, no matter how small, is capable of feeling. Of course, animals of a larger mass can have their sensatory capacity more developed.

A living entity, no matter how large or complex it may be, is composed from organized masses, for example, cells. So the entity is a conglomeration and tight solidarity of such minute living entities, and each one of them has a capacity of sensation. Every part of a single living entity experiences pleasure and displeasure in various degrees based on its capacity and complexity.

Let's imagine that some massive animal is separated into cells and each of them is placed into some medium where it can continue to develop. So, will not the result be some quantity of independent identical or at least similar animals having a capacity of sensation, in its own manner, both pleasant and unpleasant?

We see here a similarity—analogy—with the highly organized community. It is as though one whole, but you can divide it into members. Without the support of society, they will perish, but in an artificial medium they will continue a barbaric life, but having full sensation, meaning, little-persons without a connection between them. This conclusion follows if we can grasp the concept that when a person perishes a quantity of new little-persons is made from this mass. We cannot do this in reality, but it is possible. There is a lot we cannot do, but this does not mean that it is impossible. Our little-persons will have a small brain, few capabilities, little memory and imagination and etc. But each of them will have a capacity of sensation, although weak.

The question arises. Where is the limit to the smallness of a mass of a living entity that still has a capacity of sensation? A one-cell organism is very small, but we cannot deprive it of having a capacity of sensation of comfort and discomfort, although also very small.

Here is the uninterrupted chain of entities of diverse masses, from the mass of a whale, or the larger entities of other planets, to the almost invisible, ultramicroscopic bacteria, of which all we know is what we see them doing.

Can we deny them the capacity of sensation, even the smallest of any of these organized masses? Can we deny them a great magnitude of sensation, meaning, that one living entity has a capacity of sensation that is a million, trillion, decillion times weaker than another? To deny them completely of sensation, to recognize them as being a mathematical zero, is not preferable.

What occurs is that organized dead matter proceeds after organized living matter, and afterward it is more or less dissembled and unorganized. In the final end all trends toward the sole beginning—hydrogen, or to the more simple elements out of which the entire universe is composed. This is an uninterruptible chain. The 90 known simple compounds are formed from hydrogen. All rocks, minerals, gases and fluids are formed out of them and subsequently, all living entities, the simple and complex.

A living entity is only a conglomeration of other entities that are more simple, for example, cells. And a cell is only a conglomeration of complex lifeless materials, and every lifeless material is the conglomeration of a selection of these 90 molecules, gases and fluids, and these latter are a conglomeration of hydrogen atoms, and the hydrogen atom is a conglomeration of known elements of nature.

Where is the true beginning of life? Where is the initial or primordial citizen of the universe? Of course, this is an atom or its more primordial particle, and it is not known whether this is an electron or the atom's ether. We can only conditional call it a particle of ether.

Let's recognize conditionally a particle of ether as the rudimentary basis of the universe; this will be its primordial citizen. The union of such citizens creates an atom of hydrogen and other simple compounds. Now we have a community or more complex places—stages or points—of life. The union of these stages provides even more complex particles—molecules—of organic and inorganic form. Finally, the union of these latter forms evolve into living entities: from the simplest bacteria—the proto-bacteria—to the human and his perfected posterity and the residents of other worlds. These are all unions evolving from the primordial citizens, that is, particles of ether.

The collapse of this union, meaning the death of the living organism, is only the disintegration of the union, the dissociation of its members, that no longer accompanies the citizen, that is, the atoms, due to death. After the disintegration of the community of atoms, each of them can live separately and independently or else enter into a new community union, that is, into the composition of a molecule or bacteria, or even as far as some kind of other living entity that is a perfected or mature resident of other planets.

The primordial citizen is eternal. A particle of ether is the source of matter. It, essentially, is indestructible, since it is indivisible due to its being a particle of a single composition. Beyond this we have an atom of hydrogen. Many of the 90 rudimentary molecules composed of hydrogen existed for billions of years with any disintegration. The complex molecules also can exist long, for example, the molecules of alcohol, sugar, starch and others. But what is not a prolonged union is living organic matter. In general, the simpler and minor the living entity, the more its life is stable and enduring.

Bacteria only reproduce by division, and they do not die under favorable conditions of nourishment. The more solid living entities possess more diversity in their lengths of life, and the construction of the living entity does has an effect. The relationship between extent of life and mass is still not entirely explained by science.

So what do we see? The union dissolves—death, but again appears—birth. Disintegration does not destroy the citizen, but it continues to experience life, but only a more simpler—primitive—one, until it enters into a new union, that is, until it comprises a part of some kind of living entity: part of a brain, liver, muscle, and etc, in, for example, a person. It receives the sensation of life that is compatible to the cell into which the atom entered. So the citizen—now a person—from the dissolved kingdom now enters as a member of a new kingdom and experiences sensation compatible to the place it now occupies.

It is only in the brain that the living entity begins a genuine life, worthy of this appellation and our sampling of it. To reside in the balance of unions is near to non-existence, since it cannot retain time, and so on this account it has no ability to proceed with an actual life. So plant life, such as trees and grass, as though reside in non-existence.

The series of conscious life in the brain repeats an innumerable quantity of times in the mature creatures of the universe, blending into one life that is complete and endless.

The Other more Rarefied Matter, Other Worlds, Other Entities.

Science came to the conclusion that all suns, planets, all plants and animals are composed out of hydrogen. But hydrogen itself is recognized as being complex, consisting of electrons and protons. Inconveniently a strikingly resilient ether protrudes from it with its indecipherable small atomic particles. Relative to them, electrons are gigantic.

All of this indicates the complexity of all matter, the complexity of the atoms known to us. We will prove that this complexity is endless, that every atom known to us is divisible, that is, composed of parts.

Indeed, time is endless forward and backward in its boundless quantity. Matter is arranged over the course of time. If it were not this way, then we would not have our compounds of the 90 molecules known to us.

Will this modification end at some time? It can oscillate, but in general it must proceed forward. Its oscillation consists in the molecules periodically arranging and dispersing, but in general it is arrangement that primarily occurs, although exceptionally slowly.

So in time the simple atoms will increase with a quantity of 300, 400, or even 100,000 electrons and protons, and even more. Such substances will be less resilient and more solid. It is the more solid suns and populated planets, and the more solid plants and animals, that are created from them.

There is no limit to future time. After a decillion decillion years, perhaps even after a decillion decillion stages, such solid heavenly bodies and substances will form so we in comparison with them can easily be accepted as material spirits as a result of our substance being almost intangible or incorporeal. (So air and gases during eras of ignorance were recognized as spirit, as something incorporeal.)

But will we exist at this time? It is possible we will. Not everything progresses, not everything proceeds forward, not everything sharply changes. We can take as an example the organic world of Earth. A million years transpire, but not all living entities evolve into humans. Some remained behind, others took a different route. Some stopped at a very low level of development. Such were the known and unknown bacteria.

So can we also solidify in our development, that is, stop at a point where solid living entities and the relatively incorporeal can coexist and live.

In this manner, in the unimaginable distant future there will simultaneously exist not only two categories of living entities, but an innumerable quantity of them. Any one of these categories will be almost immaterial relative to those that evolved later and crudely material relative to those that evolved earlier. We, those created of a substance, as it presently seems to us to be from a very solid matter, will surface as the lightest, the optimum incorporeal living entities.

This is what a review of the unlimited future provides us, a review of the series of eras infinitely distant one from another.

Now we will consider the past. We realize it is just as unlimited as the future. Let's imagine an era distant from the present by a decillion years and by a decillion stages of evolution. At that era, the particles—molecules—were simpler, substances were less complex, less dense and more resilient. Living organisms were created from these lifeless heavenly bodies, although they were much lighter. In comparison with ours they retained a minor amount of matter and were so rarefied that they could be called incorporeal or even spirit.

The question arises. Did they disappear or do they still exist at present? It is possible that they exist just as bacteria exist contemporary with people.

Going further back, we arrive at the conviction of the existence of worlds with organisms that are even less dense. They, in comparison with the preceding, are almost nothing—when compared to their materiality, but relative to us they are the square root of our materiality. Traveling further into the past, we incur new classes of living entities retaining less and less matter. In short, from the past we have received what we will in the future, only that the series is in terms of density.

The endless future is still not accessible to us, but the past must leave its traces. And even if we are not around to encounter the more dense substances of the infinitely future eras, and not become almost incorporeal—relative to them—and conditionally spirit, then taking the past into consideration, our deliberations are no longer a fantasy, not some prospect. This exists, and we are surrounded by innumerable regiments of living entities, each one of them being incorporeal, relative to all the subsequent, and crudely material, relative to all the previous.

And matter develops, evolves, not uniformly, not consistently. A large quantity of types of matter exists simultaneously. We no longer speak of the

90 elementary molecules, from hydrogen to uranium, as we still have ether, whose density is so small that science is even inclined to deny its existence.

If matter exists in diverse forms, from almost the immaterial ether to the remarkably dense substances, as those accumulated in the center of suns, then why should innumerable regiments of living entities of the past eras not presently and simultaneously exist? And there are many types and variations of ether as well as unseen heavenly bodies. They likewise compose regiments. We cannot see these early suns and planets or the living entities that lived on them, as they were previous to us and into the distant past.

So what is our world, accessible for research by our sensations and science? It stands not at the end of time, not at its beginning. It is somewhere in the middle, and on each side of it are infinite tails of time. It will also stand in the middle, no matter how long it will live. No effect of time will change its middle position. Endlessness will always be behind it and in front of it. At the same time, an infinite series of almost incorporeal living entities will always live with us, that is, with our world.

Supporting Facts

No matter how logical and natural everything stated here may be, it would however be interesting to confirm all of this with facts or resolve the question of the intensity of their influence—these theoretical effects—on our human life.

Something extraordinary has always occurred to someone due to what was not explainable by a narrow scientific view. History accumulated quite a few of such incidences. A large number of contemporary people, those that are trustworthy, direct our attention to them, and gather and describe similar incidences in books.

The majority of such incidences, and not quite 100%, can be counted as the result of ignorance, manipulation, mental illness, forgetfulness, wild dreams that were accepted as real, blatant lies, self-deception, misunderstandings, and limited physiological knowledge.

Earlier I also thought and was convinced that the entirety of 100% fell into this sphere. However, I believed in the existence of higher planetary species of intellectual creatures similar to humans, and I suspected the existence of organisms that are infinitely lighter than ourselves. In short, I believed in living entities that were superior and more perfect, but I did not feel that they were interwoven, or at least not at the present, into the earthly affairs of people. But a refutation of this occurred to me on May 31, 1928, an event that I described in my treatise, *Will of the Universe*.

Something similar occurred to me about 40 years ago. But time has caused me to lose some of the clarity of my impression of the vision I saw at the time.[1]

We boldly speak of what we see ourselves. We have no right to vouch for others, but we are obliged to speak of our personal impressions.

It is impossible for me to not believe myself. From that moment I started to think that perhaps not all 100% of the extraordinary apparitions pertained to the sphere of deception. Perhaps, some insignificant part had some truth in it, and which I explained in this book from a purely material and scientific point of view, from the point of view of the evolution of matter.

If I saw this, then why could other people not have seen something like this also, and people no less credible than myself? If we portray ourselves completely and always credible in regards to our sensations, then what will occur with science founded on the witness of sensations that are ascertained by others?

We know that every new discovery is accompanied by the distrust of scholars, and not to mention the masses. They did not believe in meteorites, not believe in the phonograph, rotation of Earth or its spherical shape, or sunspots, or Saturn's rings. It is impossible to enumerate all the examples of humanity's unbeliefs. It is also impossible to enumerate the examples of gullibility as a result of deception. There were many more of them than disbelief in truth. We can only say that belief, as well as unbelief, were both not always justified, and that there were less changes to error in not believing, than in believing.

It is better to have a critical attitude toward everything, to over and over ascertain all sensations and means, and to accept it as factual truth only after confirmation of the discovery. Allow it to contradict our convictions, our reason, but fact remains fact and will reveal our flaws, our narrow thinking and the deficiency of the knowledge and concepts we have so far acquired.

In time science can expand and from its side confirm and explain more discoveries made earlier that we still do not understand.

Organization of the Unseen Worlds, Their Life

So we divide the worlds into regiments separated one from another by endless periods of time. It is understandable that these worlds are little accessible and one knows little about any of the others.

The final, supremely dense, visible and tangible world is the universe, with its millions of billions of suns and an even more number of planets and their moons, accessible only through exact science. Organisms that are more

[1] This was his vision of a cross-shaped cloud changing its shape into a human form, described in his autobiographical composition, *Fate, Destiny, Fatalism.*

mature than humans populate them, and they are more perfected, but yet have forms similar to animals.

How many geniuses have surfaced among us during various eras, those who moved Earth's humanity along its route to knowledge and happiness? At any time during Earth's life, such extraordinary people, valuable to Earth, will be found. How many of them are forgotten by people's ignorance? How many were never recognized and perished, not having opportunity to unveil their benevolent qualities? The future order of Earth will estrange this misfortune and immeasurable loss for humanity. Perfected individuals especially will be useful at the head of rulership.

The longer that Earth survives, the more perfected will be the available selection, and likewise the entire population will be superior. We at present are unable to imagine the superiority that all of them will reach in time. What will the persons elevated higher over the superior be like?

Earth has not reached maturity. Humanity still has many millions of years ahead of it to wait before attaining maturity. The majority of planets have matured and retain a perfected breed, and they are governing the more perfected living entities.

Every planet that has a solar system, and that has the capacity to support a population, will evolve its population into a superior quality. The more abundant the population, the more perfected the community pattern and the more superior the selected living entities and the more superior the members of the population. Gradually all the best qualities of the center of the solar system will transfer to them.

Know that what is being accomplished or what was already accomplished in our dense world will likewise be accomplished in other worlds, those that are unseen. Perfected organisms of their own species likewise reside there, along with their unions and the selected superiors from among the superiors. And their perfection and potency is unimaginable.

And so we, dense entities, are surrounded by regiments of not just dense, although perfected and potent entities, but also by regiments of ethereal entities, whose quantity is infinite, just as time of the past is infinite. Each of these regiments are ethereal relative to those of the subsequent, and crudely dense relative to all those of the previous.

What regiments of the ethereal entities have the greater influence upon us: those closer or those further? Of course it is the closer. It is the less dense that have little influence.

What is this influence? How complex is the cosmos? This is difficult to imagine. Even our limited mind is compelled more and more to increase this complexity, for whatever reason.

So immense eras will provide the possibility for this evolution into the form of spirits, although it is a rarefied materiality. My mind cannot comprehend any other. To accept some other interpretation means to reject the unity or simplicity of the universe. You arrive unwillingly to an ecstasy from the diversity in the universe that awaits us; a glance into the entities similar to us, but only perfected, sufficient and happy; an incarnation and life as spirits of an innumerable quantity of categories dispersed over eras and densities.

What type of life this will be! What diversity! What density! What amounts of knowledge! What happiness is hidden in them!

This arrangement and wealth of impressions pertains not only to humans and those entities similar to them in density, but to every atom or any part of it, to every union, to every corporeal and incorporeal substance.

The limitlessness of time not only provides us assurance for our appearance in these types of less dense entities, but also in more dense, more complex and assuredly, more perfected and wealthier sensations. But this is absolutely in the future. All the same, the future can provide us incarnation in entities lighter than those we are in at present, which are very heavy.

Appearance at any one time occurs in one density, yet during that specific epoch, entities are capable of various stages of perfection.

Morality and Ethic of Earth and Heaven

The ethical standard of the cosmos, that is, of the cognizant entities, consists in not having any suffering anywhere; not for the mature, not for any of the other immature or those creatures just beginning their development.

This is an expression of pure introversion or egoism. So if no pain or discomfort will exist in the universe, then not even one atom will arrived into an imperfect perpetrator or criminal organism. In short, the primordial citizen of the universe, that is, the atom, will then not be able to settle into a bad entity, because they will no longer be.

But we saw that the living worlds divide into two groups: one is the greater and it is populated with perfect living entities. The other is a billion fold smaller, similar to Earth, consisting of entities that are immature, but subject to hope.

It is only the benevolent that is supported in the perfected worlds. Every inclination of it toward malevolence or imposition of suffering is diligently resolved. By what method? With a means of goodness. The malevolent or those inclined toward malicious conduct remain without posterity. This means will not impose the least amount of suffering, since the breeding instinct and passions are removed. Only love can flow toward all cognizant creation, and which evolves from true altruism. It is expressed in activities that estrange suffering or the reasons for their evolution.

The potency of the perfected entities penetrates all the planets, to every area possible of life and beyond. Without imposing pain it will terminate imperfect life forms and replace them with their own mature species. It is similar to a gardener who removes unwanted plants from his plot of ground and allows only the best vegetables to grow. The primary activity of the perfected entities will consist in this along with the implementation of morality.

However a negligible number of planets will always seem to remain along with their imperfect living entities with the expectation of a beautiful and eventually supplement of perfected beings. These are embryonic planets, that is, those whose entities are subject to pains of self-generation, pains of development, like for example the present Earth and its entities. It is obvious that the destiny of some suffering cannot be avoided.

So what is the morality of these planets that are like Earth? The somewhat little growth of comprehension on our planet resulted in only the human. We can just talk about his morality, since a morality of the lower animals does not exist, their conduct is based on instinct with no cognizance of error or concept of right or wrong.

The morality of Earth is the same as that of the heavens: the estrangement of every type of suffering. This goal indicates intellect. There will not be suffering around me and I will not subject anyone to this during this life or during the endless future life. Its initial stages will assuredly occur on Earth, since the atoms of substances that have existed billions of years are tied to Earth by the force of gravity.

Human morality is complex and demands an immense education. We indicated its bases and can provide here a few comments on it.

First of all, what is necessary for all working citizens is freedom of speech, press and assembly, in general, all of such activities that are not accompanied by coercion of other persons. This presupposes freedom for the worker in every respect, so he is never under the authority of a taskmaster.

Freedom is restricted with the criminal who utilizes coercion. Their freedom and promulgation must be limited in order to deliver the population from the malevolence they commit. However, vengeance or punishment will categorically not be imposed.

Freedom is possible only when every person—except the criminal—possesses an independent means of life without interference from people. To accomplish this, every person must have the right to land, work, credit, and something that he is inclined to do and what will provide him all that is necessary for life. He must be protected from coercion and delivered from destitution.

This begins with a struggle of convictions. Initially strife will occur and many deceptions will be disseminated, but then truth will prevail, because it is stronger. Truth will indicate the best community arrangement. It consists in the best portion of humanity to rule Earth, so that each will occupy his respective or suitable spot to utilize his abilities on behalf of others.

The government by the best people, the superior representatives of humanity will provide it unity. This unity will deliver the nations from war and other forms of self-destruction or weakening, direct them to a uniform alphabet and language, teach every citizen and provide him knowledge relative to his intellectual capabilities. The government will secure his prosperity and make everyone happy.

Their inherent nature will motivate them to select what is best and attain an unimaginable mental and moral supremacy. They will gradually be delivered from criminal elements still residing in humanity. They will also be delivered from the influences of imperfect entities, but it seems, not suddenly, but gradually. No suffering of any type will exist in any place, and nothing having a cognizant nature will be subject to any noticeable pain.

No anguish from death will occur, and neither will there be any murder, deficiency, lust, hunger, thirst, cold, jealousy, envy, destruction and fear. Fear of natural death will banish from the profound cognizance of nature and will show with visual clarity that death does not exist, but what does exist is an uninterrupted cognizance and blessed existence.

1930

Socialist Philosophy

Regulations for the Society, Their Advantages and Flaws

The Right of a Person to His Individual Life

This right exists among all cultures of all nations. The murderer is arrested by the society, brought to trial and endures the punishment or the restriction of his freedom. The lower the race in the list of humanity's nations, the weaker this right, the less defined and the less implemented. It is does not exist among the lowest races. People are often treated like animals: they are hunted, killed and eaten like prize game. So the further into the past that we delve into history, so this right in society is less. Undoubtedly there were eras in each of any culture or nation when this right did not at all exist among them. It arose from need, from an embryonic stage, to its present view. But even at present, among cultured nations, this right is often fictitious. Military obligation removes this right to life. Illness, weakness, handicap, orphanage, deprive a person of what is indispensable and subsequently his right to live. Some strive to estrange this cause of death or early death, but for the time being, few attain the goal. The ideal society must banish all causes of ruin.

Is it possible for a society to exist if this right to individual life does not exist? A herd or flock of domestic animals co-exists for the most part, not mutually destructive, although they have no system of convicting murderers, no defined cohesion, no intervention of one on behalf of the other. A cow does not kill a cow, not so much because of their friendliness as much as their inability to do so, and likewise due to the uselessness of such an action.

Every domestic animal possesses an instinct that impedes it from entering into an altercation with its equal, not to mention that it does not possess a weapon with which to kill. If it were otherwise, then the herd would not exist. Natural selection caused such an instinct to develop. There is nothing more devastating than a struggle between two creatures who have about equal strength and ability. In reality, such a struggle must end sadly for the winner and even more sadly for the loser. One of the two may end up dead, while the other is wounded or mutilated, yet this can also result in death later. It is not compassion, not kindness, and not pity that restrains them from such an insane struggle, but it is rather instinct. It can be exchanged with a sensation of friendship toward an equal or something similar, with sympathy or some other sensation, but we cannot justly call this love. Virtue or kindness must be extended to every creature, and even the weakest, since gregarious animals will often inadvertently kill some other animal of another species, or other animals that are weaker, ill or wounded. This indicates that the primary means of development of this instinct is through recognizing the disadvantage of struggle against another of equal ability. Based on this premise, it seems the conclusion is that a person is guided not by virtue or kindness, but by something else. In reality, this exploitation of domestic animals, any ruthless or cruel killing of them, does not incline in the direction of kindness and justice.

Perhaps such an animal instinct existed among primitive humanity and contributed to avoiding his self-destruction. There was no justice system, but they did have instinct, although this often did not stop him from killing children, the maimed, weak, ill, old and wounded. Animals of the same species, for the most part, have equal ability and are equally armed, such as horns and teeth. But people who are armed artificially are far from having the same ability. As a result, their inequality in strength provides one side an advantage in the struggle, the side with the greater weapons. Likewise the capability of people to form regiments would also cause inequality in strength, and so the struggle is to the advantage of the regiments with the greater weaponry. The conclusion is that no animal is so inclined toward killing as is the human.

When humans—as with animals—did not possess any manufactured weapons, there was no deliberate altercations between members of the society without some firmly expressed right to individual life. Now imagine the contemporary person possessing some strongly developed morality, having a worry over the littlest impending animosity, with thoughts of a piece of bread, a home, and security for himself and his associates. And now all of a sudden, this person whose focus is the future, has no right to life! Today, such a person can be killed for the sake of food, for the sake of his property, wife or children. Can't this person just work, have contentment, and have some happiness? Won't he go crazy from the fear, from imagining that he might be refused life at some early stage! As it is, illness, death, old age and suffering threatens each person.

But this is not all. A person does not know when his life will naturally end, if he is not already old and weak. If you are 70 years of age, you can still live another ten; if you are 80, then you can live another six, and etc. Without the right to life, one that is firmly secured by those whose responsibility it is to implement the laws, humanity would be in a horrible situation. It would be hard to perceive a worse condition than this. Yet, sadly, a person seems to reconcile himself with this. In reality, militarism eternally threatens every person by forcing him to be subjected to violence and be killed or at least maimed.

Let those who want to war do so.

However, if a society forces its members to be part of a military campaign, then it violates the right to life. If the military is volunteer, then there is no violation of this right. But what we notice is that the population is deceived with good pay and promises, and so a military is assembled. The members of a society must be free to choose whether to enlist or not. Nevertheless, a society cannot survive without soldiers. The conclusion is that a military in a perfect society does not kill, but only limits the freedom of a criminal.

Now we will turn our attention to the shortcomings in the existing laws of the right to individual survival. In the most cultured countries, the criminal is still sentenced to death for committing some cruel murder. The fear of punishment restrains many people from committing a crime, especially murder. Such a penalty provides security and a tranquil existence, a hope for a peaceful life. But it is only persons who have a strong sense of morality and a fear of penalty who restrain themselves from committing a murder. The criminal seems to have very little of this to start. Often the criminal commits a crime due to a some bursting anger, jealousy or some other passion that he has inherited from his animal ancestors or from his human ancestors who previously were or presently are infected with criminal behavior. In short, the inclination toward criminal behavior is always inherited or inbred due to the conditions of early life. Sometimes it is due to the effect of unusual circumstances.

So how is a person to be executed? He did not give birth to himself. He did not create all the conditions of his existence. Would not his penalty be the same as vengeance, that is, the satisfaction of some distant bad feelings?

Occasionally the conviction of an innocent person does occur. So how is this judicial error to be rectified? We cannot return life to someone executed.

From the other side, isn't it possible to utilize a process of striking fear into potential criminals in order to estrange them from the possibility of committing a crime? It is good that people who have a sense of morality will refrain from killing. But fear of penalty is a poor means of causing goodness, so what should be done if a person is so bad, and striking fear only works on people's psyche who already have a sense of morality? If a person changes, then there is no need to strike fear into him. But meanwhile for him to ignore the law is unacceptable. So we will abandon the fear of execution and replace it with something else that is less of a formidable terror: the deprivation of freedom and other deprivations, but not the deprivation of life. Is it really possible for a person to guarantee that neither he nor any of his associates will ever kill someone? Some unexpected insult or discredit of a person or his associate can cause a person to lose his balance and cause an unpremeditated murder. Who can be secure from such a thing? Banning execution, discarding it completely from the sphere of punishments, we will at least calm all of humanity. People, associates and criminals will sigh in relief.

But are executions carried out only for murder? Are not those considered political criminals—where the thoughts of a person disagree with the thoughts of the dominant party, or due to a difference in convictions, or due to political party struggles, who are essentially wrong on both sides—also summoned to capital punishment during periods of revolution? Are you actually convinced that you will not fall into this turmoil and not be deprived of your life over a difference in opinion? Have not even virtuous persons, heroes, benefactors of humanity, geniuses, saviors, inventors and scientists been executed? And if this happens, then no executions should happen at all. No person can warrant a justifiable execution since too many innocent persons have been executed and continue to be executed. How many, for example, virtuous persons, mentally ill, great thinkers, and etc., did the inquisitions kill? So let's face the facts, capital punishment should not exist at all. Murders will still occur and the criminal must be punished, only not with an additional murder, but with a deprivation of freedom and with the fear of other penalties. Then the rest of innocent humanity will be more content, and such a kindness will surpass by many times the possibility of the increase in crimes.

The criminal may change, be reborn and become a good and beneficial person. Can any person guarantee that this is unthinkable? And if this can be, then let us leave some hope for the worst convict. This hope will color his life, provide him strength for rehabilitation, improvement, to win this struggle against himself. He will know, he must know, that everything can turn around for him as long as he gains the victory over the badness residing in him. The removal of such a cruel law will result in the contentment of humanity, in an increased lifespan in general.

Killing occurs during a struggle between two or more people. In this case there is not just one criminal, but several. Fights can occur between two persons, between two groups, between two societies, and even between two nations. In any case, those involved in a fight or war must be subject to trial for their crimes.

If I attack you with the purpose of killing you, or do some other type of violence against you, in such a situation, you may defend yourself and even unintentionally kill me in the process, and then the court will still determine this justifiable. The one who is guilty is the one who attacked you, and if he survives and is only wounded, he would be subject to trial and some type of disciplinary punishment. In war between groups, the side that was attacked would be innocent, regarding of the number that are killed in defending themselves. The same with international wars.

So how can we escape these terrors, or at least perhaps the loss of innocent lives? I codify this in my treatise *In a Perfect Societal Structure*. Fights among a few can be anticipated in a small society living in tight quarters with close associations, and they are settled in court. Likewise attacks and violence can be anticipated. But all of this is in the circumstances of a small society, and so any attempted crime can be stopped right at the root or in its embryonic stage. The same situation can be rectified if a stronger society has intentions of attacking a small society. War between nations needs to be restrained, and if it does happen, those participating must be brought to justice by the balance of humanity, that is, a union of all the countries of the world.

If a struggle is inevitable, and which is accompanied by killing and violence, then it is necessary, to whatever extent possible, to neutralize such wars and combat. To do this, laws must be introduced: prohibit the manufacture of any weapon that can be used to kill a person. The countries, societies and individuals who depart from this law will be brought to trial. A union of all nations will then not be able to war among themselves, likewise with a union of nations of other planets, but for the meanwhile this is not a threat to us. Such a union can demand that its members subject themselves to their authority. Wars will be terminated, but some weapons will remain, only they will not be specialized, such as automatics. It will be impossible to demilitarize immediately, but eventually war will be reduced to almost a fistfight and will end quickly and without the number of sacrifices that occur with the use of sophisticated and highly destructive weapons, or improved models presently in the planning stage for future use.

We see that a person's right to life materializes through contemporary institutions that are far from perfect. This right will increase if we introduce these supplements to the existing laws.

> Capital punishment is always and in all situations abolished (although the commandment not to kill is ancient, we need to now make it concrete).
>
> Manufacturing any weapons to be used to destroy a person is strictly prohibited by law and so punishable.
>
> Fighting between persons and societies is subject to judicial proceedings by a special organization of people.

Maybe some will tell us that war serves a good purpose in order to select the strongest and most enduring. Yes, but only if it is accomplished within the arena of a fistfight. But we do not have such wars at present. Even then, is a fistfight between the best person on each side the best way? What also occurs at present is an artificial selection.[1] So wars are not fought in any perfected manner, and nations need to accept the fact that not one of them is any better than another, including the nations whose populations are greater. But an international union with highly intelligent individuals will restrain animosity.

It seems that the contemporary establishments, if compared to zero, are still good and must be observed until they are replaced by something better. Laws naturally progress toward perfection, and the existing institutions played an important role and were better than complete anarchic crime. The execution of a murderer was better than complete impunity or vengeance taken by the offended relatives of the murdered. All of these regular and often unjust and erroneous executions did improve the general condition of humanity. It caused less people to be inclined toward murder. Even a non-punishable fight between persons to the death brought some benefit, since the winners were people who were stronger, more agile, accountable and vigilant. An individual improvement in humanity occurred in this manner.

Development can be accomplished in two manners: individual and societal. In the properly oriented society, one and the other must occur simultaneously, because one without the other will lead to a regrettable conclusion. This will promote the unification of groups, their solidarity, friendship, love among themselves, and create a tight circle of people. Initially the individual strength of a person developed, then it was the strength of the family, later it was the small society, and then the large societies and states. Now we are at the point of needing unification of all

[1] Referring to military conscription.

countries. But not all at once. Humanity has a history. Gradual development is what has been inevitable. But no one will interfere in increasing the speed of our development and leading it to its summit, to the unification of all the populations of the world. We already spoke of the right to life, no matter how imperfect the society. Masses of peoples for various reasons are deprived of the indispensable opportunity to life and so die or terminate their own life early. The societal arrangement should be such that this early demise should not exist. Only then will this right attain perfection.

The Right to Property, Land, Animals

This does not exist among animals. What we see here is that might is right. The predatory animal takes advantage of the life of the weaker animals. He appropriates from them what he finds necessary for his purpose. The herd does not share itself, but vicious animals take what they want. The stronger, bolder, braver is the one that acquires the most and often all. It is only with an abundance of food, like grass for example, when there is no struggle. Very seldom is there a struggle between the very strong and very weak, because the weak will use its instinct and either hide to save himself and not become lunch for the stronger, or else be caught. The stronger has less to risk and so simply devours the weaker.

So it was at one time in humanity's environment. There were no laws, no such thing as personal property, but might was right, or what they called their natural right and which extended to other people. Even an animal recognizes the right of personal property when pertaining to its mate and offspring. A person likewise, no matter on what level he stands, he cannot but recognize even to some insignificant degree this right toward his own family. Without this, the propagation of the human race would be impossible.

Nevertheless, the existence of lower forms of life without personal property or family rights is possible. Such for example are the single-cell organisms, many insects, and etc. Their species propagates immensely. The species of humans and animals that have not recognized family rights or personal property must eventually extinguish or eventually demise by natural means. Weak and hungry children perish before their full development. So the family rights of personal property, just as with the family right of life, was generated at the very lowest stages of the existence of animals and humans. Then it extended to a small society, further to a larger one, and then to an entire government. Where the property rights were weak in some family or society, then struggles would arise to appropriate them from the other, the weaker family or society. Whichever was the weakest of all was eventually destroyed by them. So the societies that recognized the right of private property persisted, while the others disappeared. But this did not yet expand to all of humanity. A strong state or their union can indeed utilize these rights in the ways it seems fit.

Of what does the right of personal property consist? The first person who discovers something in nature, he has the right to utilize it. Whoever manufactures or modifies something, he has the majority right to it. Spouses do not take food that they find away from each other, and likewise they do not take it away from their offspring. In this manner, the family right of personal property is established, although not without many and coarse deviations. Some cannibals even eat their own children without any regret. A female spider will eat its male spouse, and some male animals will eat the female's offspring as soon as they are born. Such a right among the stronger has extended to captured or weaker domestic animals. Among humans this right has extended to captives, subjects, or the weaker members of the human race, and even to members of a person's own family. And so the rights of one family are often violated by another, stronger family.

The union of two families that are not feuding, that have a cohesion, and can recognize mutually the rights of personal property, provide a doubly strong union and subsequently [develop] the ability to violate the rights of individual families. The union of three families wins over dual-family unions. So small societies appeared, each possessing a mutual agreement to recognize their right of personal property and proceeded with the intention of a victory over the weaker unions. This victory could not have come about if there was no preliminary recognition of the right of personal property by several families. In reality, if this was not established, then the families would have fought among themselves and could not have created a victorious union. Over the course of time, the number of such united societies grew, and at the same time the size of societies grew, since the societies with the larger population subjected and absorbed those with a lesser population.

A person's right to property consists in rights to the land, the production of an individual's personal labor, to things, to animals, to people, and the right to bequeath, after death, in a will, all of your property to your children, relatives, or other people, as you see fit. More or less all of these rights flourish in most countries. We will proceed to discuss the positive and negative sides of these rights.

The Right to Land

At the present time, every person—under normal conditions and in most cultured countries—can become the owner of some quantity of land. To what does this lead? The owner of a large expanse of land is usually a capitalist. He is often a very talented person, unless he acquired the land through some inheritance. Capital provides him the possibility to well organize the utilization of the land, import the best of machinery, implement the best methods of agronomy, and as a result, with a small number of workers and with a comparatively small expense of labor and capital, attain brilliant results, that is, to harvest from the land up to 10 times more produce than with some obsolete means of farming.

Let's imagine an ideal situation. If we were to suppose that this person, and one having a high sense of morality, decides to not bequeath his land to unworthy persons, those who will live excessively extravagantly, and who will not personally hire a large number of workers to work the land. Then in this ideal situation, intense cultivation of the land is more advantageous for the community. But does it always occur this way? Often an heir of limited ability inherits the lands, one that is not even interested in farming. So he rents the land to some other farmers and the intense cultivation of the land is converted to a weak. If the owner should himself organize cultivation of the land, then his effort will be moderate and limited by his ability and he will not received any brilliant results. So little is to be gained by either course of action.

From the one side, an owner who is morally weak receives rent money and this causes him to devolve into lethargy, since he is unable to effectively work or meditate on the situation. So he uses the income by squandering it on his passions, something that people—even the strongest—seem to do so easily in such situations. So we destroy the landowner. In addition to this, others will notice the free income of the landowner and will surround him like a bunch of cannibals and he will squander his free income on them. This temptation is great and we often see such predicaments in life.

Weak owners of land, united into an organized cooperative, can introduce public roads and high quality equipment, a general cultivation, and operated by a person knowing the business and who is known by them. The result of this course is the same as that of the ideal capitalist venture. What is difficult is the education of the community on taking this course. Pretty often such organizations do not succeed and the matter turns for the worst, than if the land was cultivated on an individual basis.

Furthermore, weak landowners with successful ventures, assigning the administration to capable hands, can also delve into inactivity once receiving regular income. All of this is far from an encouraging picture. So the members of this owners' union receive income, doing nothing. How can this be? The land belongs to the people; no one else can own it.

The ideal right to land consists in each person owning almost 25 acres of land. How is this calculated? The surface area of Earth's dry land will be divided into 1.6 billion equally-valued parcels and distributed.[1] Of course, they would be of various shapes, and each parcel would be assigned to one living person. In summary, each human soul would have the right to almost 25 acres of cultivatable land. A family of ten has the right to almost 250 acres of land.

This is the ideal and which obviously humanity will not accomplish quickly. But such an ideal is still far from complete. We cannot even start to think of accomplishing something like this unless a general association is already implemented. Although all existing laws regarding land ownership need to be observed, it is now necessary to gradually modify them to draw them to the ideal. There cannot be any of these irrational class struggles, futile suffering and self-destruction that seem to be inevitable with every disastrous upheaval. We can, for example, legislate laws requiring owners of large tracts of land every year to return a certain percentage of their land ownership. We can also limit the amount of land inherited. They can also turn over the greater portion of their land yearly to the rest of humanity. Some can also be compensated in one way or another for the loss of their land.

The right to land ownership is a great step forward. Animals do not have such a right, and among the less-cultured nations it is regularly violated and difficult to implement. Even in the cultured nations much intervention is necessary in order to guard this right, even in regard to a small parcel of land. Rights to land provided the possibility for the owner to seriously utilize the land, improve its use and cultivate it. Such a concept energizes an owner, having hope for a good harvest, to work hard: to plow, fertilize, level, water, harvest and store his produce. Agriculture would be impossible unless a right to land ownership was possible. Even then, we have not included cattle husbandry or domestic animals. If it were not for the present educated attitude toward these issues, we would not have attained even a tenth of our present population.

The conclusion we can make is that immense ownership of land, beyond a reasonable norm, is harmful to humanity and to the owners themselves. Laws must gradually limit the size of land ownership and lead toward a standard of about 25 acres of land person or some equally-valued parcel.

Based on this, however, we should not conclude that farm labor will be mandatory for every person. The land needs to be available to any person wanting to work on it, and this is on an individual basis. Society seems to have a different approach and to make this complex, but we do not have the time to discuss this more thoroughly. We must not forget that members

[1] This was Tsiolkovsky's estimate of world population at the time, year of 1919 along with his estimate of the dry land surface of Earth.

enter into a society voluntarily, but members who enter into a high society are selected.

Can someone refuse the right to land ownership? Of course, and then it will pass to some other person who will take ownership. If society as a corporate body will take sole possession, then this will agitate the individualists and cause animosity and the disasters of conflict. In developed societies, only an insignificant portion of their members actually work on farms in the cultivation of the land, since most of the labor is expended by the use of machinery. But if more were to be employed in farm labor, the time would decrease to perhaps even two hours a day per person. So do all people need to work the same amount on some parcel? The value of a person is hard to define. This question will be resolved for the most part in the perfect society. In some cases, the value of one person regarding some assignment is 100 times that of another. In such a case this person handles the management, and so is well rewarded, as opposed to the one operating equipment or doing manual labor.

The Right to Raw Materials

Suppose that on someone's private parcel of land, valuable deposits of some type of ore are discovered: coal, silica, gold, and etc. Then this parcel no longer has the same value of other parcels of the same size, and so to make it of equal value the owner would have to decrease its price by 100 or 1,000 or even a million times relative to its true value. But no populace considers that adequate; and so they will seize it from the owner and give him another property of similar size and of moderate value. Now, land needs to be considered based not just on its surface area, but what is underground, and its natural wealth causes a reevaluation of the parcel and its wealth will be used for the benefit of society. All of humanity will be benefited and individuals will receive a share. Suppose some person's allotment of land includes a waterfall: its value as a significant source of mechanical energy is very great.

Every individual or society has the right to take advantage of the raw materials of its land, until the raw material no longer causes the property to have a higher value than some standard parcel. In life we do not see this attitude toward raw materials. The owner of some land, discovering some valuable ore on it, keeps it for the most part as his own. Such a right leads to capitalism and the owner's acquisition of power without any effort from his side.

The right of inheritance, combined with the right of the possession of property, gives birth to another category of persons who are able to exist without being any benefit to humanity, and are rather a burden. But there's no need to arbitrarily dismantle the laws; instead, they should be gradually reformed. Initially 50%, meaning half, of the discovered treasure can be permitted them, and then this will decrease annually to 40%, 30%, 20%, and so forth to zero.

Prospecting would thus be encouraged. People involved in this would be rewarded with good living conditions and other comforts of life relative to their discoveries. New legislation leading to the discovery of nature's wealth will benefit all of humanity.

The Right of the Development of Raw Material, or the Right to a Person's Effort

Why this right of people is expressed is only obvious. I make something from my materials, using my own hands I make a chair, dresser drawers, house, hammer, or equipment, and they belong to me. If I please I can gift them to anybody I want, exchange them for something else, or bequeath them to whom I want. This is so natural and totally just. The more a person works, the more productive this work, the more the reward for it. It would be strange if I worked, while some lazy person, sloth or failure came along and took advantage of my work or my design, the product of my intellectual faculties. If it were this way then the diligent person would cease to work. The number of idle people would increase and all would be set on the course toward parasitism and internal ruin. If the right to raw materials exists, then even stronger is the right of a person to what he manufactures from such raw material. Such items can be even more valuable than gold or jewels that are discovered.

So let us imagine for ourselves the contemporary person and contemporary laws. One person can accomplish much. He utilizes the forces of nature and uses machines in order to produce a thousand time, a millions times, more than another who is less smart, less skilled and less diligent. Can some evil evolve from this? Our fortunate manufacturer exchanges his goods for the nourishment so necessary to him, for clothes, for gold, and other expensive goods that are easy to preserve and which will not quickly decay. He dies and all he accumulated is passed in its entirety to humanity. What a personal loss!

But perhaps the ordinary manufacturer hides or destroys his products, perhaps he gifts them to some favorite persons but who are lazy and they evolve into parasites. Perhaps he will distribute all he has to satisfy some personal motive. But what he does distribute, he also has the right to receive something in return of equal value.

The introduction of parasitism is the temptation of such a situation. How can this be? How to escape this evil: the evil of parasitism, the evil of luxury, passions, whims, the suppression of goodness and the possibility of misanthropy.

More than often a person falls into such a situation due to the temptation to satisfy passions that are far from a virtuous character and which destroy him and any good he does. Then comes the necessity of regulating the exceptional worker's rights of personal effort. So opulence, luxury and excessive passions are prohibited, and the introduction of parasitism is prohibited. A portion of the populace, regardless of their talents and due to their moral qualities, will remain outside the society and will live a life close to contemporary norms. They must possess the rights that I have described above on the organization of the society. Laws must little by little incline toward restoration and toward the regulation of the right of possession of what is produced by a person's own effort

The right of personal property has another bad side. The law shelters this right, but the primary worry of protecting personal property lies on the owner. His anxiety is proportional to the amount he has accumulated. The more he possesses, the more he worries, and the more energy he expends in the process. The moment is coming when all the energies of the possessor will be directed to and concentrated entirely on protecting his assets. But how do you protect products that decay quickly? If you sell them for gold, silver or other metals, or jewels, then the purchaser needs to find a means of securing them and that will not be cheap. Some persons may even have to make the decision whether to protect their assets at the cost of their life. True, some valuables are protected in banks and other edifices erected specifically for this purpose. But even then, can such institutions be completely trusted to protect your assets?

In America there are bandits, robbers, who are equipped with the latest weapons that can destroy steel and concrete walls and break into safes. It seems to me that it is easier for the possessor to deliver his manufactured goods or products to society, and for society to quickly distribute them, while he receives assurance from the society that he will receive proper compensation for his effort. This is considerably safer and [such a process] does exist here in an embryonic stage.

Due to necessity or desire, the possessor sometimes receives the his own products for part of his effort. But what can he personally do with such an excess? He can only eat so much and the rest can rust or decay or just spoil. Expensive clothes and living conditions generate envy and cause bitterness, and can also serve as a reason for their violent death. Yet free distribution of items and food to people just corrupts them and generates envy, animosity, distrust, falsity, parasitism between them, and often ingratitude and hatred. To help the underprivileged is a very difficult matter and not everyone has the strength to do this. Sometimes excess is invested for the purpose of making a profit, but this has its own share of problems.

Profit can be made by various methods. It can be some investment of a fixed percent over some period of time, or speculation, gambling, and other means that can consist of risk, or organized crime. But illegal means are to be quickly proscribed by law.

I only want to show that although a wealthy person has many courses he may to take regarding his benefit to society, should he take such a course, laws to regulate evil resulting from wealth are inevitable. They can gradually limit the evil, with the consensus of both the wealthy and the society, with condescension toward the ignorant and toward human passions: greed, embezzlement, megalomania, pride and etc. Wealth can also be used as a source for useful ideas. Only in rare circumstances is a thinker capable of extracting gold from his genuine ideas. If this occurs, then this source of wealth is the most honorable. But for such people, wealth is also dangerous.

The Right to an Idea, The Right to an Invention

Contemporary law provides the right to an invention, but not to an idea or concept. There is an inherent error here: an idea and an invention are the same. One and the other can be equally beneficial for people and so both must be encouraged and distinguished.

The right to an invention is not allowed every person, but only to the person who has the means to acquire one (but in general I do not have the USSR in mind) and only after much paperwork and a long interval of time. The person who discloses his invention before receiving the proper paperwork loses his right to it. This is a very disturbing injustice. The inventor announces—and often because of not knowing this law—his discovery in its entirety before it is legally codified, in order to faster introduce it into the life of society and put it into production. He only harms himself in the process, since after he's made the announcement it is often difficult to prove that the invention belongs solely to him.

A law of security[1] was introduced in part so the invention would become a secret known only to the one government, and especially if it is of a military character. The invention can make a person wealthy, but only the one who already has the means for manufacturing. So now the assignment of patents can only strengthen the accumulation of wealth into a few hands. Nonetheless the existing rights to inventions are beneficial for a person since they encourage inventions that produce immeasurable benefits to humanity.

[1] The process of acquiring copyright or patent.

It is regretful that this encouragement is insufficient for the poor and weak, that it requires much expense, and that it is more accessible to capitalists. In this manner the mass of inventors are stuck in a helpless state and their inventions fall into the hands of foreigners for some meager sum and even any honor due to the inventor slips away from him.[1]

It is extremely important to know who invented specifically what. It is to the people's advantage to know this exactly in order to encourage the inventor and his kind, to define the value of his qualities and to utilize them always.

It is not only the invention that needs to be lauded, but also every good design or concept. The old—the less, while the new—the more. Particularly the person who is considered to have brought humanity some new and great concept needs to be distinguished and recorded permanently in books, or at least a person who resurrects old and forgotten ideas that are still useful. We need to make it easier to generate new concepts in society. How to do this is already described in my other treatises. Not even one good idea, as well as its author, should be hidden or submerged in some futile location.

In this manner do I propose the right of personal property that pertains to each new useful idea and the right to the royalty on the resurrection of old ideas, forgotten ideas. The honor of their discovery must remain with them and be permanently recorded in books.

If some worthless person finds a means of fraudulently ascribing to himself great ideas and inventions, then this will just encourage liars and worthless people and this will promote the increase of their type and so lead humanity astray. This conduct is not only unjust, it is also damaging to others.

So how can we encourage ideas and their fulfillment? In a way that will promote more great ideas and more creators. First of all, an attitude of awareness of the complete surrounding environment is needed. New societies should already have this on their agenda. Further, an impartial deliberation of ideas and their promotion relative to their worth. So if an idea has some promise, all the powers of the society will be used to bring it to fruition. Its value will increase more once the idea is put into practice. Its author will be detailed in some book, with descriptions of his merits and inventions or innovations. He will be rewarded with less demanding work, improved living conditions combined with his requests, and all of this will

[1] Tsiolkovsky was in a similar predicament with the patents on his dirigible design, and wanting to sell them to foreign investors to acquire money so he could manufacture them. But nothing evolved from this.

direct his effort toward assisting others with their ideas, and which will be promoted by society.[1]

Not only will encouragement promote virtue and a satisfaction of a higher moral sense, but an abundant display of talent, abundant efforts to generate ideas, a self-denial of mundane vanities, an increase in the population of talented persons, and the generation of conditions for more geniuses to appear, will all be for our success.

The Right over People and Animals

The right to [own] people is now considered a violation and is prosecutable in all countries. Such is slavery and the sale of people. Every type of involuntary servitude and coercion of labor is restricted by law and such laws are good.

In the past, slavery and sales of people flourished. This displayed the fact that the moral qualities of the people were very low and this was exposed by only a few, those better representatives of the human race.

Then humanity's despotism took a different turn: slavery exists in the form of the authority of the capitalist and the power of the strong over the weak. Under the guise of freedom, we have slaves just as in the past and we laugh at them, as much as we want, because of their desperation to acquire a ruble. Even so, exchanging the slavery of ancient times with capitalism is a great step forward. The strong and capable have an entrance to freedom through capitalism. They work the land, they are craftsmen, salespeople; and it is difficult, but they can find access to freedom and independence. The other means, and an easier one, is to change sovereigns and so escape the slavery entirely. During earlier eras this could not be done. Presently, access to freedom consists of education, guidance, assistance and direction. Just as a shepherd leads sheep to greener pastures, so the elected humanity leads society along a path to happiness. Their right to do this is acquired from people voluntarily and cheerfully, and with a desire and hope to better their life and that of the entire human race. At the same time, average people must have rights over retarded or immature people, criminals, and children until they reach the age of maturity.

Even greater are the rights of people over animals. Presently these rights are almost unlimited, except the right to maltreat animals. Of course, we need to keep the right to destroy dangerous insects, harmful creatures and predators. But the right of humans over the higher animals must be somewhat limited, since animals are unable to intervene on their own behalf. This can be accomplished only by people.

[1] This materialized with Tsiolkovsky with his government pension in later years for his work on rocket and dirigible designs.

Indeed, the incessant and cruel butchering of millions of higher animals solely for their meat cannot continue eternally. This humiliates humanity. Is it not possible for us to stop consuming meat as food, and so stop the killing, and have a vegetarian diet instead? Are bananas so bad? And fruits and vegetables, or fat derived from plants, and bread, and etc. What can vegetation not provide us? Eventually the animal kingdom must be terminated, and without harm or pain; either through a total separation of the genders or painless castration of the males. These laws should not be legislated immediately, but little by little based on the agreement of the majority of the population, as much as practically possible.

The Right to Inherit Land and Property

This right has a good side to it. An active, industrious, moderate and accountable person who accumulates property uses his brains and back, and then he dies. The children inherit this property. Is this just? Is this reasonable? Is this useful to humanity? Essentially, qualities of the father are in part bequeathed to the children. The children are now financially secure due to the inheritance and they mature, while others who do not have this financial security die from hunger and other deprivations. Of course, the children of some lazy and irresponsible person should not have the same portion as do the children of the industrious person. Allow the children of the industrious parents, those in the better situation, to grow in better conditions! So will we maintain a generation of industrious, accountable and moderate people. Humanity will then become wealthy as a result of possessing this element of diligence. The introduction of the right of inheritance of the close relatives of the deceased is an unconscious endeavor of society to perfect the human race. This is more of an artificial selection, rather than a natural.

Of course it would be easy for society to appropriate inherited property from a weak widow and helpless children. But society does not do this, but protects them from predators, and this is a step forward. So it was in ancient times when the amount of property to be left to heirs was small.

But is this contemporary method of inheritance the most reasonable and beneficial for the perfection of the human race, through artificial selection? In part, yes, but not entirely. What do we see from the heirs of property from wealthy parents? Is it always good? It is often useless and dissipation. Far from always are the good qualities of the father passed to the children, as if the father actually had good qualities to pass!

A compassionate person cannot become wealthy, since the tendency is for him to distribute his property to the hungry and needy. Of course, he might do this wisely: be very economical in his distribution, according to his surplus, and so remain wealthy. If he should distribute half his income for uses by the needy, while using the other half to further invest in his business, he will still have a surplus. But we do not see such a system in place, although it is possible and can be implemented.

Greed consumes a person. Can a person find some means of overcoming it? Yes, but altruism and generosity combined with a sense of financial accountability and economy is a great rarity. Usually, other than acquiring an intellectual inheritance, the reason wealth is increased is an immeasurable stinginess toward others as well as toward himself.

So in the majority of situations, we confirm the right of inheritance for spoiled people who are indifferent toward humanity's poor and who are often greedy.

Let us suppose that the heir acquired only the better qualities of their parents. Let's see what evolves from this. He has everything, he is flattered, everyone caters to him, they guess at what he wants, they confirm the talents he thinks he has, that he is above other people, that he has a right to luxury, to expend his passions, to the best women. But with such opulence of every type, all he does is destroy all his better qualities that he inherited from his parents. Being an essentially good person from childhood, he morally weakens due to his unrestrained passions. He loses his strength to resist them, and more and more he falls into immoral servitude. Nothing now seems to motivate him to work or think or search. His physical and intellectual strengths are now atrophied, and as a result his offspring are not much different, ending up having a regretful, useless and parasitic existence. But much is required to maintain such rich people who are worthless to society.

What is particularly required is an austere modification of the laws of inheritance by society that permits such parasites to live, doing nothing as they collect rent from property or interest income. If the heir receives a portion of an inheritance, he still needs to learn to live independently, to select some vocation that he enjoys, and to endure some deprivations, but those not harmful to his health, some loss of means of livelihood, the occasional inability to gratify all his passions, the absence of flatterers and spongers, and without unearned compliments. This will force him to direct all the strengths of his mind, soul and body to attain the fulfillment of his needs on his own. This will motivate better qualities in heirs and provide them with a good work ethic.

Monetary Issues

We will now migrate to the issue of the significance of money. It is not necessary to immediately abolish gold, silver and other materials that are a means of exchange and accumulation of wealth. We notice that a manufacturer who sells his products to society is provided a receipt, and then he provides them an invoice for payment of the products. But can't this invoice be used in the same manner as money, to purchase something else for the same amount as the invoice? Of course, it is possible. But this can also lead to a greater evil. Fraudulent receipts can be printed as well and counterfeit invoices just as easily as counterfeit money. But at the same time, this will avoid the decrease of value of printed money since the means of exchange will be the value of the original products.

Money consisting of coins made from gold or some other precious metal has its bad side, but is permitted and hopefully for just a short while longer, until something better is legislated. Paper money is twice as bad. Gold and other precious metals can still be used for productive purposes: as fillings for teeth, decorative roofs, dishes, buttons, machinery, instruments, electrical devices, all depending on the amount of gold and other precious metals available.

Science Fiction

On the Moon[1]

I awoke and, still lying under the blanket, I meditated on the dream that I just dreamt. I saw myself bathing. But since it was winter, it seemed to me especially pleasant to dream about bathing in the summer.

Time to get up!

Stretching, rising. How easy to do this! Easy to sit, easy to stand. What is happening? Is my dream continuing? I feel how easy it is for me to stand, as though immersed in water up to the neck: my feet are barely touching the floor.

But where is the water? I don't see it anywhere. I wave my hands. I do not sense any resistance. Am I still sleeping? I rub my eyes. Everything is the same.

Strange! Nonetheless, I still need to dress.

I drag a chair near me, open a closet, take out some clothes, pick up some objects and—I seem not to understand anything!

Did my strength increase? Why is the air so stuffy now? How am I able to pick up things that earlier I could not even move?

No! This is not because of my feet, or my hands, or my body! They were much heavier earlier and to move them took so much of my strength.

How did my feet and hands gain such stamina?

---

[1] This science fiction story is in the form of a person's dream when he falls into a state of delirium. The dream consists of him awakening on the moon, and his experiences there and with a friend who is a physicist. In the end he awakes from his dream when he recovers from his delirium.

Or perhaps it is some kind of power that is pulling all of these objects upward and so making my effort easier? But in this case, why is it pulling so hard? It seems to me that in just a while, I will be drawn toward the ceiling.

Why do I not walk, but jump? Something is pulling me to the side, some opposing force, pushing my muscles, coercing me to skip. I cannot resist the temptation—I jump. It seems to me that I rather slowly rise and then just as slowly come down.

I jump harder and from a sufficient height I gaze at the room. Oh, oh! I injure my head against the ceiling. The room has a high wall and I did not expect a collision. I will not be this careless anymore.

A noise. However, I just woke my friend. I see how he is turning over and getting out of bed, but somewhat skips out from his blanket. I will not start to describe his surprise, similar to mine. I saw such a spectacle when a few minutes before I experienced the same, but I did gain a large satisfaction by watching the expanded eyes, humorous poses and unnatural vitality of my friend's movements. I laughed at his strange cries, very similar to mine at the time.

I allowed my friend the physicist to get over his surprise before I turned to him with the question of resolving my puzzlement over what had happened. Did our strengths increase or was gravity reduced?

Both one and the other proposition were equally astounding, but it was not something that we could take for granted. The desire to know the reason suddenly generated in both of us. My friend, adapted to analysis, soon sorted out the episode and stunned and confused my mind.

"According to the dynamometer, or spring scales," he said, "we can measure our muscular strength and know whether it increased or not. So I lean against the wall with my feet and pull the lower hook on the dynamometer. Do you see? 12 lbs. My strength has not increased. You can try the same and convince yourself that you have not become a heavyweight or someone like Ilya Muromets.[1]

"I wisely agree with you," I expressed, "but the facts are contradictory. Explain to me how I can raise the end of this bookcase, which weighs no less than 120 lbs? Initially I justified my doing this because I thought it was empty, but opening it, I noticed that it was full. Explain likewise, why I am able to easily jump a height of 15 feet."

"You are able to lift heavy loads, jump high, and feel yourself light—not as a result of your strength increasing. This supposition was already shown to be false with the dynamometer. But it is the result of the decrease of gravitational pull, and of this you can convince yourself by means of the spring scale. We will even calculate the specific amount it has decreased."

---

1 A valiant hero and champion of Kievan Russia epic poems and folk tales.

With these words he picked up the first weighty object he could grab, which was about 30 lbs, and placed it on the dynamometer.

"Watch," he continued, looking at the arrow on the scale, "a 30-lb weight is only showing 5 lbs. This means that the force of gravity is reduced by a factor of 6." Thinking for a moment, he added, "This exact scale of gravity exists on the surface of the moon, a result of its smaller volume and the lesser density of its composition."

"So we are on the moon?" I chuckled.

"If we are on the moon," laughed the physicist, condescending into a humorous tone, "this is not so bad, since not only is such a wondrous event possible, but it can also be repeated in the opposite direction, meaning that we can return to our homes again."

"Wait! Enough of your jokes. Suppose we were to weigh some object on regular lever-arm scales. Will the reduction in gravity still be noticeable?"

"No, because the object weighed in the balance dish has likewise decreased in weight, in the same proportion as the earlier weight. The relative balance is not violated, regardless of the change in gravity."

"Yes, I understand."

Nonetheless, I decided on another course and attempted to break a stick in half. In despair I tried but did not have sufficient strength to do so. The stick was not so thick and yesterday it was weak in my hands, but now I had less strength.

"Stop this, you obstinate thing! Put it down," said my friend the physicist. "Better to think about our present situation. The entire world is exciting and filled with changes."

"You are right," I said, throwing down the stick, "I just forgot about things. I forgot about humanity's existence, and with whom I want, and you also want, to share our thoughts and feelings."

"So what has happened to our friends? Have other upheavals occurred?"

I pulled open one of the curtains that were pulled closed for the night to reduce the lunar light that interfered with our sleep. I turned to chat with my neighbor, but suddenly he moved back. What a horrible situation. And the sky was as black as the blackest lead pencil. I closed the curtain.

Where is the city? Where are the people?

This was so wild, unimaginable, when at this time the local area was supposed to be brightly illuminated by the sun! Were we somehow transferred to some desolate planet? But I only meditated on this thought, unable to say anything, and then just incoherently mumbled something.

My friend threw himself at me, supposing I had gone insane, but I directed his attention to the window, and he threw himself at it and likewise went numb. We looked around.

The curtains of all the other windows were still closed, as earlier. We clung to each other with some timidity, and then started to raise just the edge of one curtain again. We then raised all of them, and finally decided to leave the house to observe the gloomy sky and surroundings.

Despite the fact that our thoughts were consumed with the imminent stroll, we noticed something else. While we were walking about the large and high room, we had to conduct ourselves carefully due to our muscular strength, otherwise the soles of our shoes would slide along the floor uselessly. Even though we would not suffer any hurt from a fall, as would occur on wet snow or on ice frozen over the ground, since we would just bounce back, we were still cautious. When we wanted to move in strictly a horizontal direction and fast, the first thing we had to do was make a noticeable bend forward and then our movement was extremely easy—but it had to be done slowly. To descend a ladder, step by step, was so boring. And to take short steps was tedious and slow. Soon we stopped these maneuvers, good for Earth but a joke here. For movement, we learned to skip. To go up and down, we just jumped over 10 or more steps, like excited athletes. And then once we jumped right to the top of the ladder, or jumped right from the window to the ground. In short, circumstances forced us to change into skipping animals similar to grasshoppers or frogs. So, running from inside the house, we jumped outside and ran skipping in the direction of one of the nearest mountains.

The sun was blinding and seemed a bluish color. Covering my eyes to shield the sun's light and the shining reflection of the light by the surroundings, it was still possible to see stars and planets, and which also seemed to be a bluish color. But none of them sparkled, and they appeared like nails with silver heads beat into a black ceiling.

And there was the moon—its final quarter! It could not astonish us, since its diameter now seemed to be three or four times more than the diameter of how we viewed it earlier.[1]

Here it also shined brighter than daytime on Earth. Tranquility, clear weather, cloudless sky; no growth was apparent, no animals. It was a desert with a uniform black vault over it and a motionless blue sun. No lakes, no rivers, and not even one drop of water. Even though the horizon was blanched with an appearance of water, this was a mirage; in reality it was just as black as the zenith! There was no wind to rustle even a blade of grass or cause the treetops to flutter. The chirring of grasshoppers was not heard. No birds or diversely-colored butterflies could be noticed. Only mountains and mountains, awesome high mountains, the tops of which did

[1] The visitors are actually looking at Earth from the moon, but to them it still seems to be the moon, although they cannot figure its larger size.

not shine from any snow. Not one flake of snow to be found anywhere! In the distance were valleys, plains, plateaus, and all of them filled with rocks, black and white, large and small, and all of them pointed, shining, jagged, not smoothed by any winds or waters, and which were not here anyway, and which would have played with them with a joyful shout. None of this to erode their surfaces.

And here was a place completely smooth, although wavy. Not one rock was seen, only black fractures in a serpentine pattern were noticeable all over. The solid ground was rock. Not one spot of the black surface was soft. No sand and no dust.

What a gloomy picture! Even the mountains were bare, shamelessly naked, to the point that we could not see a thin veil of anything on them, not even a transparent bluish haze like what hovers over Earth's mountains and indicates a distant atmosphere. These were austere, strikingly distinctive landscapes! And the shade. They are so dark. And what sharp changes between the darkened and the lighted areas. There were no soft fog banks, as we were accustomed to and which only an atmosphere could provide. Even the Sahara Desert would appear as a paradise in comparison with what we saw here. We were desperate to see scorpions, locusts, gusts of hot sand created by dry wind, not to mention the rare occurrence of meeting some gaunt bush or clump of date palms.

We had to start thinking about returning.

The soil on the surface was cold and breathed with coldness, to the point that our feet were chilled, but the sun would bake them. In general, the sensation due to the cold was quite uncomfortable. It is like an ice-cold person warming himself in front of a stone fireplace but he cannot seem to get warm because the room is extremely cold. The heat will reach his skin but penetrate no further because it cannot overcome the coldness of the room.

On the return trip we warmed up by jumping with the agility of a deer over piles of rock that were six feet high. They were granite, porphyry, sienna, rhinestone, and various transparent and semi-transparent quartzes and silica, all of them having a volcanic origin. Then of course, we noticed a few furrows.

Now we were home! In our room we were well again; the temperature was uniform and agreeable. This motivated us to attempt new experiences and to deliberate on all that we saw and noticed. It was clear that we were located on another planet. There was no air on this planet, and not any kind of atmosphere. If there was gas, it would cause the stars to sparkle. If there was air, the sky would be blue and shades of gray would appear on the distant mountains. But how were we able to breathe and hear one another? This we could not understand. From the many scenarios we were able to acknowledge the absence of air and any gas of any kind. As a result we were not able to light a cigar, and we would just lose all our matches anyway trying to light one. A sealed rubber container was crushed without any effort, which not have occurred if there was any kind of gas inside of it. This condition, the lack of gases, does exist on the moon, as scientists have concluded.

"So are we on the moon?"

"Did you notice that from this point the sun does not seem any larger or smaller than from Earth? Such a view can only be observed from either the Earth or its satellite, since both heavenly bodies are located about the same distance from the sun. From other planets, it will seem either larger or smaller relative to the Earth. For example, from Jupiter, the sun is 5 times smaller, from Mars it is 1 ½ times smaller, but from Venus, it is the opposite, being 1 ½ times larger. On Venus the sun's heat intensity is twice what it is on Earth, while on Mars it is half the amount. So those are the differences between the two planets that are closest to Earth. On Jupiter, the heat absorbed from the sun is 25 times less than Earth. We do not see anything similar here, disregarding that we have the ability to use inclinometers and other measurement instruments."

"Yes, we are on the moon. All the evidence testifies to this. Why do you say," I mention to my friend, "that Earth and moon are the same distance from the sun? It seems to me that the difference is very considerable. As much as I know, it is equal to about 250,000 miles."

"I said almost. Since the amount of 250,000 miles only makes up 1/400 of the total distance to the sun," the physicist expressed. "Just 1/400 can be ignored. I am now tired, and not so much physically, but mentally. I can feel sleep wanting to overcome me. So what does the clock say? We got up at 6:00, and now it is 5:00. This means 11 hours have passed. Meanwhile, judging by the movement of the shadow, the sun has almost not moved. You can see the shadow of that steep mountain and that it has hardly moved toward our house. You can also notice the shadow of the wind vane on that rock, and it has barely moved. This is new evidence that we are on the moon. In reality, its rotation about its axis is so gradual. Here daytime lasts about the same as our 15 days, or 360 hours, and nighttime is the same length. This is not very comfortable. The sun interferes with sleeping. I remember what I experienced when I had the opportunity to spend several weeks during one summer in the [northern] polar regions. The sun did not set below the horizon and this horribly annoyed us. But there is a big difference between here and there. Here the sun moves slowly, but in the same order; there it moves fast and every 24 hours the sun changes its location above the horizon. But what we can do here is what we did there: close the shutters."

"But isn't the clock correct? Why is there such a disagreement between my pocket watch and the pendulum clock on the wall? My pocket watch says it is 5:00; but the one on the wall is behind at only 10:00. Which one of them is right? Why is the pendulum rocking so slowly? It is obvious that this clock is slowly falling behind."

"Pocket watches cannot lie, because their motion is dependent on a tightly-wound spring, so it is the same whether on Earth or on the moon, as opposed to a pendulum that swings, whose period depends on gravity. We can compare this with something similar, such as our pulse. I have a pulse of 70. Now it is 75. A little faster, but this can be explained by nervous excitement, dependent on unusual conditions and strong reactions. Nonetheless, there is still another possibility to ascertain the time: at night we will see an Earth that completes a rotation in 24 hours. This is the best and infallible clock."

Disregarding our slumber that seemed to be overwhelming us, my physicist could not tolerate the incorrect time on the walk clock and fixed it. I noticed how he removed the long pendulum, precisely measured it and shortened it by a factor of 6. Now the precious clock turned into a ticking machine, but it was not so bad, because the short pendulum still swung slowing, although not as slow as when it was long. The result of this metamorphosis was the consistency in time measurement between my pocket watch and the wall clock.

Finally we lay down on our cots and covered ourselves with light blankets, but it seemed stuffy in the room. We almost did not even need pillows or our mattresses. In fact, we could almost have slept directly on wooden boards. I could not sever myself from the thought that it was still too early to go to bed. It was the sun and the time. But everything was quiet, just as with the entire lunar landscape. My friend stopped responding to my questions and I quickly fell asleep.

I awoke a happy person. Vigilant and with the appetite of a wolf. Up to this time the upheavals had deprived us of a normal routine of dining. I want to drink. I remove the cork. What happens? The water is boiling. Sluggishly but still boiling. I hold the decanter with my hand and worry about getting burnt. No, the water is just warm. I cannot drink it this way.

"My physicist, what do you have to say?"

"Here it is an absolute vacuum. The water boils because there is no atmospheric pressure. Let it boil, do not put the cork back. In a vacuum the boiling will end when it freezes. But we will not let it get to the point of freezing. This is enough! Pour the water into a cup and then put the cork back in, otherwise more will just boil away."

Liquids pour slowly on the moon! The water in the decanter settled, while in the cup it continued a slow turbulence, but the longer I waited, the slower it was. Then what was left in the cup turned to ice, and the ice slowly evaporated until nothing remained.

So now what are we supposed to eat?

Bread and some other more or less solid food can be easily consumed, although it quickly dried out since the house was not hermitically sealed. The bread quickly turned into rock; the fruit shriveled and shrank and likewise became somewhat hard. Nonetheless, their peels still retained some moisture.

"Oh, boy. This habit of mine of eating hot food. Now what will I do? I cannot in any manner start a fire in this place. There is no wood, no coal, and even a match will not light."

"Can't we use the sun to do this? I heard that eggs can be baked in the burning sand of the Sahara!"

So we modified the pots and bowls and other kitchenware so that the lids were flat and held their tops on tightly. All of them were filled with what we had available for our culinary taste, and they were set all together in a sunny spot. Then we gathered all the mirrors in the house and placed them in a way that would allow the reflection of the sun to fall on the pots and bowls. Not even an hour passed before we were able to eat well cooked and hot food.

One thing was not going well and so we really needed to hurry. This will make sense if I explain to you that the soup boiled and cooled not only on our plates, but even in the throat, esophagus and stomach. I have to admit that more than once we shoved food into our mouths and swallowed it fast. As soon as I started to chew, guess what, instead of soup I had a chunk of ice. It was amazing how our stomachs were able to adapt; the lack of air pressure extended the time it took to swallow something.

In any case, we were full and sufficiently content. We did not understand how we were living without air, and how we ourselves, our house, yard, garden and kitchenware and food and liquids and what we had stored in the cellar and barns, were able to be relocated from Earth to the moon. Then we started to have doubts and we thought: maybe this is a dream, or a hallucination, or perhaps we were in a demonic trance? However we accustomed ourselves to our situation and dealt with it in part due to curiosity, in part due to indifference. What we could not explain did not surprise us, but any danger of dying from hunger while all by ourselves and in such an unfortunate state—never entered our minds.

How can I explain such an impossible optimism? You will discover this as I unfold our adventures. We wanted to go for a walk after dining. I was not able to sleep anyway and I feared having a stroke. I urged my friend to go with me.

We are standing at the edge of a wide yard, in whose center was a tall rock, while a fence was erected along the sides with other buildings. Why was this rock here? Someone could hurt himself on it. The soil in the yard was just ordinary dirt, soft. The two of us with our combined strength were able to lift this rock weighing about 150 lbs and toss it over the fence. We heard how it silently struck the rocky surface of the moon, although we actually heard nothing. It was the vibration of the ground that reached us, and not sound waves. This is the way we were able to hear any impacts that we caused on the moon's surface.

"So how are we able to hear one another?"

"Hardly. The sound here should not travel the way it does in an atmosphere."

The ease of such movement motivated us with a stronger desire to crawl and jump. What a happy time was our childhood! I remember how we climbed on the tops of roofs and trees, pretending we were cats and birds. This was fun. But now we were able to jump across fences and gullies, and run to win a prize. I gave myself passionately to this. But later I needed to remember my age, since I had little strength remaining in my hands. I was able to jump and run decently, but it was difficult for me to climb a rope or a pole. I dreamt about having great physical strength; I would repay all my enemies and would reward all my friends. Child and champion are one and the same. Now these dreams of strong muscles were just a joke. Nonetheless, my desires, fervent in childhood, materialized here. Thanks to the negligible lunar force of gravity, my strength increased six-fold.

Other than this, there was now no need for me to utilize my entire body in order to increase my strength. So now a fence was no obstacle for me and no more than a threshold or a stool that on Earth I could easily jump over and without having to run. And now in order to verify this preconception of myself, we stood and then jumped and we flew right over a barricade. We even jumped over a shed, but to do this we had to run. But still how pleasant it was to run, it was as though I could not even sense my feet. Then I sped and went into a gallop.

With each stride, every time my heel hit the ground, we traversed several yards, and especially in a horizontal direction. In one minute we went around the entire yard, a distance of about 1000 yards at the speed of a race horse.

We took measurements: when galloping at a light pace, we rose 12 feet every step we took, while horizontally we traversed 30 feet or more, depending on our speed. To the gymnasium! Awesome! From a height of 24 feet I dropped and barely felt any pressure on my heel. I should warn my friend about this, but I maliciously urge him to jump. Raising my head, I call to him, "Jump! It is nothing. You will not get hurt."

"You urge me in vain. I know for sure that this vertical distance on Earth is the same as about six feet high. I have no doubt there will be a little pain in my heels." My friend flies downward, a slow descent, especially at the start. In all it takes him about five seconds, enough of an interval to meditate on a number of items.

"Well now, my physicist."

"My heart is beating fast. Nothing other than that."

"To the garden. Let's climb some trees. Run through the trails."[1]

"So why have the leaves there not withered?"

Fresh growth; protection from the sun; tall linden and birch trees. Like squirrels we jumped and climbed along the shallow branches and they did not break, but it was not as though we were fat turkeys. We slithered over the bushes and between the trees and our movement reminded us of flying. Oh, this was pleasant. How easy it was for us to keep our balance. Hanging on a branch, ready to fall, but the inclination to fall is so weak and even any tendency to imbalance is so slow that the smallest movement of the hand or foot is sufficient to readjust ourselves.

The open space! The immense yard and garden seem to be a cage. Initially we run on level spots, then we encounter holes but not very deep, and up to 60 feet wide. With a running start we fly over them like birds. So now the ascent begins. First it is shallow and then steeper and steeper. What an upward slope. I fear getting short-winded, but there is no reason to do so. We climb without obstructions, longer and faster steps up the slope. The mountain is high, and even with the light gravity of the moon we are still exhausted. We sit. Why it is so soft here? Why are these rocks so pliable? I take a large rock and strike it against another. Sparks fly.

"We have rested. Let's go back."

"How far from our house?"

"Not far. About a thousand feet."

"Can you throw a rock that far?"

"I don't know, but I will try."

We each took a medium-sized jagged rock. Who will toss it further? My rock flew over our house and by quite a distance. But watching it, I thought for sure it would break a window.

"And yours? It went further."

[1] The entire yard with plants, trees, fences, and animals, were also transferred with the house to the moon in his dream.

We then figured a shot from a weapon would be interesting. A bullet or cannonball must fly horizontally and vertically at least 70 miles.

"But will gunpowder work here?"

The explosive element in a vacuum must display itself with even greater power than in an atmosphere, such that the latter only obstructs its velocity. As far as oxygen is concerned, there is no need for it, because as much as necessary for ignition is already contained in the power.

We arrived home.

I said, "I will spread some gunpowder on the window sill with the sun shining on it. Focus a magnifying glass onto it and you will see the flame, bursting, although noiselessly. The familiar smell will be noticeable for a moment and then it will disappear."

"You can fire something. Do not forget to install a percussion cap."

"We will orient the weapon in a vertical direction so that the bullet will be nearby when it returns after the shot."

Fire, a slight sound, a light agitation of the soil.

"So where is the pad?" I cried. "It should be right here, near us, even though it is not smoking."

"The pad ejected together with the bullet and will not be far behind it, since it is only Earth's atmosphere that interferes with it following after the lead bullet. Here the lint pad flies high with such a velocity, just like the bullet. Take a feather sticking out from a pillow, and I will take a pig-iron ball. You will be able to throw your feather and strike the target, even at a distance, just as easily as I can with a ball. With a certain amount of effort I can throw the ball about a thousand feet, and you can throw your feather the same distance. True, you will not kill anybody with it, and you will not even feel any effort on your part when you throw it. Let's throw our metal balls and feathers with all our strength, each one of us separately since we have different abilities, at the same target. Over there is red granite. We will see how the feather will travel somewhat faster than the iron ball, as if carried by a strong wind."

"But what is going on? It has been three minutes since we fired the shot and the bullet has not dropped," I said.

"Wait two more minutes and it will surely return," answered the physicist.

Then exactly after the indicated interval of time we felt a light agitation of the ground and saw a bouncing pad not far from us.

"Where is the bullet? Shouldn't the impact of the fall cause some agitation?" I asked.

"Definitely. Due to the heat of the discharge, the bullet shattered and the small fragments flew in various directions."

Looking around we indeed found some small fragments, obviously parts of the falling bullet.

"This bullet sure did fly a long while. So how high did it rise?" I asked.

"I guess about 50 miles. This height is the result of the small force of gravity and the lack of atmospheric resistance."

My body was exhausted and it needed rest. The moon was the moon, but still I was feeling the impact of the long jumps I did. We did not always land directly on our feet during our extended flights, and the result was that sometimes we did hurt ourselves, even in the low gravity.

In the course of 46 seconds of an ascent, we were not only able to view the surroundings from a sufficient altitude, but were able to move our arms and legs somewhat. Initially we were not able to rotate in space. Then later we were able to train ourselves to do forward and reverse somersaults, and in such situations we tumbled in space: first it was once, and then up to three times. Such movements were an interesting experience, and interesting to view it from the side, as I did the movements of my physicist who performed this without any trouble at all, and so distant from the ground. To describe it all I would need an entire book.

We slept for eight hours. The region became warmer. The sun rose higher and baked even the most incremental crevasses of the body's surface, but the ground also warmed and no longer exuded any coolness. In general, the effects of the sun were such that the ground warmed and was even somewhat hot.

It was now time to take measures of caution, since it became apparent to us that by the time it was noon, we would be cooking or burning. So what could we do? We had some plans.

"For a few days we could live in the cellar, but it will be about 250 hours before night finally arrives, and the cellar is not deep enough for the day heat not to penetrate there. Other than that, we will miss every comfort of our home due to their absence while we are in this secluded space."

"I suppose that it is far better to tolerate discomfort and boredom than to be cooked."

"Wouldn't it be better to choose a cave that is deep? Let's move there and spend the rest of the days and some of the nights in pleasant coolness."

"This is considerably more enjoyable and poetic. But I still prefer the cellar."

"Desperation for the necessities of life will drive a person insane in such a place."

"So the cave it is. The hotter the sun blazes, the further we will descend. Nevertheless, just a few yards deeper will be sufficient for us."

"Let's take umbrellas, and take provisions in sealed boxes and barrels. We will throw some coats over our shoulders to cover ourselves if we need to go in the sun and warm up, if it should get too cold."

A few hours passed, and during this time we were able to dine, rest and further discuss lunar gymnastics in the absence of gravity, and what wonders Earth's acrobats would be able to accomplish here.

But we could not postpone it any longer; the heat became infernal. It was so extreme that the outside areas that were illuminated, such as the rocks on the ground, were so hot that we had to tie thick wooden boards to the bottom of our shoes to act as insulation. In our hurry, we dropped glass and ceramic dishes, but they did not break, so weak was the gravity.

I almost forgot to mention the fate of our horse that we had brought here with us. This unfortunate animal, when we wanted to harness it to the carriage, somehow tore out of our hands. He initially rushed away faster than the wind, turning somersaults and hurting himself, not comprehending the power of his inertia. Unable to stop soon enough when he reached a pile of rocks in his gallop, he ran right into it and broke apart. The meat and blood splashed all over the place, froze, and then dried.

By the way, I want to say something about flies. Here they cannot fly, but only jump, and each time about half a yard.

So, we gathered everything we could not live without, with an immense load on our shoulders, which seemed to us to be quite amusing, since due to the weak gravity all that we were carrying felt almost empty and light. We closed the doors, windows and shutters of the house, so it would be less affected by the high temperature, and departed to find a suitable cave.

While we were searching for stuff to take, what struck us was the sudden change in temperature: the places that had long been lit up by the sun radiated heat like a violently burning fireplace. We tried to avoid these areas and refreshed ourselves and rested in any shade we could find, even under a rock or under a cliff. But these places were unreliable since the sun would move and heat that area regardless, so there was no more shade and coolness. We grasped this fact and so sought a cliff where, although the sun would shine, it would only be for a short while and not long enough to heat the rocks.

We found a cliff with walls that were almost vertical. Looking down we only saw the beginning of the wall: it was black and seemed to be fathomless. We walked further and found a slanting descent leading, as it appeared, to hell itself. Some of the steps down were fine, but then the darkness thickened and now in front of us nothing was visible. To go any further seemed dangerous, risky. We remembered that we had grabbed an electric light—since a candle or lamp would be useless. The flashlight turned on and momentarily illuminated the cliff for about 100 feet downward. The descent appeared to be feasible.

We are now at the edge of a bottomless pit, we have found hell. The desolation mesmerized us. Its darkness, first, is explained by it lying in a shadow, and due to its narrowness and depth, the rays from the illuminated surroundings and high mountains did not penetrate into it. Second, there was no light from the surface atmosphere, as we have on Earth, where even at the bottom of a well, light penetrates down and such darkness is not incurred.

We grabbed the sides of the cliff as we went down. To the extent that we descended, so did the temperature decrease, but it never got below 60° F. Apparently that was the mean temperature of the space in which we were located. We selected a comfortable, level area and spread our coats on the ground and tried to make it as comfortable as possible for us.

So what's going on now? Is it nighttime? Turning off our flashlight, we gazed at the dark, heavenly turf and the innumerable stars shining rather brightly over our heads. However, our chronometer indicated that very little time had passed, and the sun in no way could have suddenly set.

Now this! A clumsy movement and I dropped the flashlight and it broke. But it was just the glass, while the filament continued to burn, since it was in a vacuum. If this had occurred on Earth, it would have gone out, burning itself out in the atmosphere.

Out of curiosity I decide to touch the filament. It breaks and everything goes dark. We do not see one another; the only thing visible is the uppermost edge of the cliff, and the black ceiling seen through this long narrow tunnel is now illuminated with even more stars.

I cannot accept that daytime at the surface is as hot as we figured. I can no longer tolerate being here and so with difficulty I find another flashlight we took with us, turn it on and start upwards. The ascent is brighter and warming. Then the light blinds me and the flashlight is useless. Yes, it is daytime, and the sun and shade difference is still here. It is hot! I need to go back fast.

With nothing else to do, we slept like badgers, but we could not get warm in our burrow.

On occasion we would leave the area, find some shady spot and observe the course of the sun, stars, planets and our large moon[1] that, in relative comparison with the moon as seen from the Earth, to us it was an apple, while from the Earth it was a cherry. The sun moved almost equally with the stars and it was hardly noticeable that its circuit was slower than theirs, nothing like what is noticeable on Earth.

[1] They are actually looking at Earth from the moon.

Midday was approaching. The period of shade was getting shorter. The moon had the appearance of a narrow scythe, becoming paler and paler as the sun got closer to it. It was an eclipse. On the moon it comprises a regular and grandiose display; on Earth is it rare and insignificant: the speck of a shadow, not quite as large as the head of needle (and sometimes several miles long, but this is also a needle's head in comparison to the size of Earth), inscribes a band on the planet, traversing in a pleasant manner from city to city and residing in each of them a few minutes. Here the eclipse covers either the entire moon, or in most cases a significant part of its surface, such that the complete darkness continues several hours.

The crescent appeared and in line with the sun it was hardly noticeable.

The crescent was no longer visible.

We climbed out of the crevasses and looked at the sun through a dark lens.

Now only half the sun was visible.

Finally the eclipse's last part disappeared and everything was immersed in darkness. An immense shadow overwhelmed and covered us.

But the blindness quickly disappeared. We see the moon and a number of stars.

Yes, I see the light of the evening twilight, which we enjoyed watching on Earth. And the surrounding edges were flooded with scarlet, deep as blood. Thousands of Earthlings gaze with their naked eyes and through telescopes at us, observing a complete dark [new] moon.

Eyes of my friends and relatives, do you see us?

While we were feeling sorry for ourselves, a red tiara developed uniformly and beautifully. It was hovering over the entire region of the moon: this was the middle of the eclipse. Here on the side opposite from where the sun was located, the edge paled and then it shone. Then it started to shine even more and appeared like a diamond set in a red ring. The diamond evolved into a part of the sun and the tiara disappeared. The night turned into day and our catalepsy slowly faded as we thought about what had just occurred. We enthusiastically started talking.

I said, "Good for us for selecting this shady place to make these observations."

For the sake of convenience, and in part for physical exercise, we took note of a number of rocks in our cavern. We decided to collect the ones that had not had time to get warm, as many as we possibly could, and drag them outside and arrange them in the entrance to serve as a means of protecting ourselves from the heat.

No sooner said than done.

In this manner we were always able to climb to the top and sit alongside the pile of rocks and successfully make observations. Then the rocks started to get hot. We could drag new ones there, since deep in the cavern there were plenty, and with our excessive strength—relative to the moon's gravity—it would be easy for us. This we did after the eclipse was over.

Other than this work, immediately after the eclipse we started to measure the latitude of this region of the moon where we were located. It was not difficult to do, taking into consideration the time that elapsed during the recent eclipse and the height and location of the sun. In this way we calculated our latitude as 40° N, as opposed to our original assumption that we were exactly on the equator.

Then midday passed, the time since the sun rose, which is the same as seven Earth days. However the chronometer indicated that the time of our residency on the moon so far was equal to five Earth days. We concluded that we had arrived on the moon early in the morning, meaning its morning. This explains why, when we awoke, we found the ground still very cold; it had not had time to warm up, being horribly cold after the preceding prolonged night of 15 Earth days.

We slept and woke and every time we saw above us more and more new stars. This was the same familiar pattern as we would see from Earth, all the same stars, only that the narrow hole in which we resided did not permit us to immediately see most of their quantity. It was as if they sparkled on a black background and their motion was 28 times slower.

We recognized Jupiter in the distance and could distinguish its satellites with the naked eye, and we also watched their eclipses. Then Jupiter disappeared and the North Star appeared. Pathetically, here it does not play such an important role.

We need to stop sleeping so much! We decided to devise plans.

"At night we will leave the cave, not immediately after sunset while the ground is still baking hot at a high temperature, but after a few hours pass. We will visit our home and see what is going on there, whether the sun's rays have done any damage. Watching the Earth from here is like watching the moon from Earth: it has its phases which we watched with imaginative curiosity. From our region at midday, it is the new moon, or should it be better described as a new-Earth. At sunrise it is the first quarter; at midnight it is similar to a full moon; at sunset it is like the final quarter.

This is not bad, but only as long as we reside in the hemisphere that is visible from Earth. But as soon as we pass into another hemisphere, one not seen from the Earth, then immediately we are deprived of some nighttime illumination. This is the unfortunate and hidden hemisphere, hidden from Earth, since the Earth never faces it, and this is very intriguing for many scholars. It is unfortunate because its residents, should there be any, are deprived of a nighttime lamp and a majestic scene.[1]

Really, are there residents on the moon? What are they like? Are they similar to us? To this time we have not met them, and it was hard to meet them anyway as we just sat for the most part in one spot and occupied ourselves more with gymnastics than with selenography. Especially interesting is that unknown half.

The black skies here at night are eternally covered with masses of stars, the greater part of which are vaguely seen and require a telescope. Even then, their weak radiance is not dissipated or diffracted by several layers of atmosphere, and is not stifled by interfering light, as on Earth.

There are no especially deep recesses here where gaseous fluids can accumulate or where a lunar population can reside. These are the types of conversations we had in order to pass the time while awaiting sunset and the night. We even awaited it with impatience. But it was not very lonely; we did not forget about an experiment with wood oil that the physicist attempted earlier. The matter is that we succeeded in acquiring drops of a immense size. So drops of oil attained a size of an apple, although they flowed out of the trees at a slower rate than on Earth under the same conditions.

We still did not forget about our alimentary canal. Every six to ten hours we would nourish ourselves with food and drink. We had brought a samovar[2] with us and we often drank Chinese herbal tea. Of course we could not light it in the normal fashion, since we had no atmosphere in order to ignite coals or small twigs, so we just brought it out into the sun and filled it with small blazing rocks. It heated rapidly, but we did not allow it to boil. Even then the heated water gushed out of the faucet and caused a great deal of steam to flash, since there was no atmosphere or gravity to stop it. Such tea was not very pleasant, since overheating it caused the water to spray or splatter in every directly.

What we did was to put the tea in the samovar and allow it to heat to a high temperature, then we removed the hot rocks and waited for it to cool. Finally, we drank the prepared tea, not burning our lips. Even then, this relatively cool tea still burst out of our glasses and weakly boiled in our mouths, like seltzer water.

Soon it was sunset.

We watched as the sun touched the summit of one mountain. On Earth we would watch this display with our naked eyes. Here this was impossible since there is no atmosphere or water vapor, and as a result the sun in no way lost any of its blueness or its heat or radiant strength. A person could

[1] Referring to not seeing the Earth from the moon's backside.
[2] A traditionally Russian, special design kettle to boil water.

only glance at it quickly without the use of dark glasses. It is nothing like a crimson and weak sun on Earth during sunrise and sunset.

It descended and sank, but slowly. A half an hour passed after the moment it touched the horizon and yet half of it was still not hidden.

In Petersburg or Moscow a sunset takes only three to five minutes. In the tropical regions it is only two minutes. It is only at the poles that sunset can take several hours.[1]

Finally the last fragment of the sun terminated its illumination by hiding behind the mountains. But there was no twilight. In place of twilight we see surrounding us a rather bright radiance that projected from the summit of the mountain range and other elevated parts of the distance. Such light was enough that we were not plunged into darkness over the course of many hours. One distant pinnacle shone like a light bulb over the course of 30 hours, but eventually it also went out. The stars and the reflection of the light off Earth—like the moon to a person on Earth—is all we had to break the darkness, even though the stars were more of a display than a source of light.

Let's return to our view of it. Earth from the moon—although it appears to us like a moon—has a surface area 15 times larger than the moon, and as I mentioned before, it is like an apple compared to a cherry. The amount of light it reflects is 50 to 60 times more than the amount of light reflected by the moon when seen from Earth, and so a person could actually read under such circumstances and it did not even seem like a night, but a surreal day. Its radiance, without a special screen, did not allow us to see much of the starry sky.

What a view! Welcome, Earth! Our hearts beat tumultuously. Memories of the place penetrated our soul. How dear Earth now became to us, which we earlier reviled and debased. We see it as a picture covered by blue glass. This glass is the atmospheric ocean of Earth. We see Africa and parts of Asia, the Sahara and Gobi deserts, Arabia! The regions that are waterless with a cloudless sky. No blemishes whatsoever on it and always in plain view of any lunar residents. Formless white puffs were its clouds, while desolate areas appeared as a dirty-yellow or dirty-green.

The seas and oceans were dark, but their shades varied, depending obviously on the extent of their waves or calm. Bits of the water's surface were white, no doubt reflecting the foam at the tip of each wave. Its waters were also covered by clouds, but not all the clouds were white as snow; the rest were gray. No doubt they were covered by higher levels of clouds composed of frozen crystallized dust. The two diametrically opposite ends of the planet were especially bright: these were the polar snows and ice. The northern whiteness was cleaner and had a larger surface than the southern.

[1] Toward the winter and summer solstices.

If the clouds were not in motion, it would be difficult to distinguish them from snow. Nonetheless, the snow for the most part lay deeper in the atmospheric ocean, and so the blue color that covered it was darker than the same blue over the clouds. Some snow-capped ranges were noticeable scattered all over the planet and even along the equator, but others were more striking.

We gazed at the brilliance of the Alps! The summits of the Caucasus. The mountain ranges of the Himalayas. The patches of snow were more consistent than the clouds. Using the telescope we were able to verify many details of the surface, and how we enjoyed doing this!

It was time to eat. But before descending into the cave, we decided to discover how hot the ground was at the time. Departing from the rock barricade we had recently built, and which was already rearranged several times, we realized we had just walked into an intolerably heated bathhouse. The fiery heat still penetrated through the mountainside. We returned backwards; it would be a long while before the ground would cool.

We eat in the cave, whose entrance is now no longer lit up, but an awesome number of stars are visible. After every two or three hours we would walk out and observe what we have for a moon, which is actually our Earth. From some places the clouds were obstinate and refused to move and this made us impatient, although at the same time we enjoyed watching them, until the redness of a nightfall affected the view of Earth's surface.

The next five days we just hid in the moon's viscera, and if we did walk out, it was locally and just for a short time.

At the end of the five earth-days, which brought us to almost the middle of the night in moon-days, the ground was somewhat cooled and we decided to attempt a journey across the moon, meaning through its valleys and mountains. Up to now we had not traveled through any depressed regions.

These darkened, immense and depressed regions of the moon were assigned the name "seas," although this is completely improper, since there is no trace of water. Will we possibly find traces of Neptunian effects in these seas and other even more depressed regions, traces of water, atmosphere and organic life, which in the opinion of some scholars have long disappeared from the moon? There is a supposition that all of this did exist at some time, even though at present it cannot be found in the ravines and depressions: that water and atmosphere did exist, but were adsorbed or embedded over the course of centuries into its soil, combining chemically with it; that there were organisms, some kind of plant growth of a simple nature, some types of shellfish, because where water and atmosphere exist, there is also algae, and algae is the start of organic life, at least the lower forms.

As far as my physicist-friend was concerned, what he thinks and uses for a basis is that the moon never at all had life or water or atmosphere at any time. Even if there was water, even if there was an atmosphere, the temperature was so high that no organism of any kind could possible survive.

May the reader forgive me for providing you my physicist-friend's personal view, which possesses no proof anyway.

When we finish our round-the-world journey,[1] then we will see who is right. So grabbing our luggage, which is now rather light, since we ate and drank a lot of it, we leave our hospitable and comfortable cave and in the darkness of the moon's night, we journey back to our residence, which we soon find.

The wooden posts and other parts of the house and other out-buildings, made of the same material, subject to the continual effect of the sun, crumbled and were all charred on the surface. In the yard we found the remnants of some barrels that were ruined. We had left the corks in them and allowed the barrels to remain in the direct effects of the sun's heat, so the expansion of water vapor inside them eventually burst them. Of course, there were no more traces of water; it all evaporated without a trace. On the porch we found some fragments of glass from the lamp, the frame made of some thin metal. It disintegrated and the glass just fell out. In our house we found our goods less damaged. The thick stone walls were intact. Everything was secure in the cellar.

Gathering necessary items from the cellar so we won't die from hunger or thirst, we departed on our protracted journey to the moon's pole[2] and then to the opposite secretive hemisphere that has not been yet seen by any person.

"Shouldn't we follow the course of the sun toward the west, inclining a little toward one of the poles?" the physicist proposed. "Then we will be able to kill two rabbits:[3] the first rabbit is attaining the pole and edge of the hemisphere; the second rabbit is to escape the intense cold. Following behind the sun, the cold areas will still be warm should the sun get ahead of us, and the temperature will always be moderate. This would be best as we get closer to the pole, that the average temperature will get lower."

"Yes, of course. I guess we can do that," I stated, wondering at the strange theory of the physicist.

"It is very possible," he answered. "Just take into consideration the lightness of running on the moon and the slow movement of the visible sun. Indeed, the maximum circumference of the moon is about 6,700 miles. We need to run this distance in 30 days, so as not to get behind the sun, which is about 700 hours in Earth-time. The result is that we need to run about 10.5 miles an hour."

[1] Meaning, around the moon.
[2] He is referring to the South Pole of the moon.
[3] The Russian equivalent of killing two birds with one stone.

"Run 13 miles an hour on the moon!" I shouted. "Why, that's just trivial."

"Well, you understand this is easy."

"Even taking it easy, we should be able to do twice as much," I continued, remembering our mutual gymnastic exercises. "And this will allow us to sleep for 12 hours every day."

"Other parallels are that the closer we are to the pole, and since we are traveling westward, the less we will have to run to catch up to the sun," the physicist explained. "We can gradually travel with less speed. However the cold of the polar regions will not allow us to do this. The closer we get to the pole, in order not to freeze, the closer to the sun we need to get and move to the regions, although polar, that receive light from the sun for a more prolonged period. The polar sun stands at a shallow point above the horizon, and so its heating ability is almost incomparably weaker, and so at sunset the ground is hardly warm. The closer we are to the pole, the closer we need to be to sunset, in order to maintain a constant moderate temperature."

"To the west! To the west!"

We slid like a shadow, as though hallucinating, silently touching the pleasantly warm ground with our feet. We almost circumvented the moon and the sun's light was shining bright, presenting the region as a mesmerizing picture, covered with a blue glaze whose thickness grew at the edges the closer we came and the less the darkness. At its further reaches we could not distinguish dry areas or water or any cloud clusters.

Now we see the opposite hemisphere [of the Earth], a region rich with deserts. But after 12 hours it will be the opposite: rich with water—one large Pacific Ocean. It reflected the sun's rays weakly, and so if it were not for clouds and ice floes that brightly reflect, the moon [on this side] would not be as bright as it is now.

We easily run to high ground and even more easily descend from it. Occasionally do we submerge into a shadow, where more stars are visible. For the meanwhile, we run into only low hills. But the highest mountains do not pose an obstacle to us, since here the temperature of the place is not dependent on its elevation. The summits of mountains are just as warm and free of snow as are the lower valleys. The uneven plains, cliffs and ravines on the moon are not scary. We jump directly across the uneven areas and ravines, even those 40 to 60 feet wide. And if they are too wide and inaccessible, then we endeavor to run the periphery or else use a pole to help us jump over the narrow gorges.

I need to remind you of our little gravity, and so we have no need for ropes, as you must understand.

"Why don't we run to the equator? We haven't been there yet," I remarked.

"Nothing stands in the way of us running there," the physicist agreed. And so we immediately changed our course.

We ran extremely fast and so the ground seemed to be somewhat warmer. Finally our speedy stride became impossible due to the heat and so we ended in a spot that was more warmed by the sun.

"What will happen if we were to run, disregarding any heat, at this speed and in this direction toward the west?" I asked.

"In about seven Earth-days at this rate of running, we would see the initial rays of the sun over the summits of mountains, and then the sun itself rising in the west."

"So is the sun supposed to rise where it ordinarily goes down?" I expressed my doubt.

"Yes, this is correct, and this display will prove itself to be true before our eyes."

"So what will happen? Will the sun show itself momentarily and then hide, or will it rise as usual?"

"As long as the speed we are running is over 10.5 miles an hour, and this would be along the equator, then the sun will be moving from west to east, where it will eventually set, since our ground speed exceeds the moon's rotation. Running long enough, we will catch up and the sun will seem to rise from the west. But then all we need to do is stop, and then it will proceed in its normal manner and then submerge below the horizon."

"But suppose we were to run at exactly the same speed as the moon's rotation, about 11 miles an hours, what would happen then?" again I asked.

"Then the sun would appear to stand still in the sky as it did at the time of Joshua the son of Nun, and everything would stay the same."

"Is it possible to do that on Earth?" I added.

"It is possible, but only if you are in a condition to run, drive or fly on Earth at the rate of about 1,040 miles an hour or a little more [at the equator]."

"How can this be done? That's 15 time faster than any hurricane. Something more than I can handle."

"Well, well. It is possible here, even easy, while over there on Earth it is completely unthinkable," said the physicist, pointing at the Earth.

So did we deliberate, sitting on rocks, since we could not run any more due to the heat, which I mentioned above. Exhausted, we soon fell asleep.

A noticeable freshness woke us. We vigilantly arose and jumped about 15 feet, again running toward the west, in the direction of the equator.

You remember that we had determined the latitude of our residence to be 40° north, so that a considerable distance remained to reach the equator, even not counting that a degree of width on the moon is less distance than on Earth, the sizes of the two being comparable as a cherry is to an apple. One degree of latitude is only about nine miles on the moon, while on Earth it is about 70 miles.

As we drew nearer to the equator we, among other things, convinced ourselves that the temperature of the deep ravines actually increased and attained a high of 145° F, but not any higher. Then it started to decrease, meaning that we were moving to another—the lower—hemisphere. Our exact location we determined using astronomical methods. But before crossing the equator, we traversed many mountains and dry seas.

The beautiful formations of the moon's mountains are known to Earth's residents. For the most part they are round with a concave depression inside. The dish is not always empty, and is not always a recent crater. Often in its center another entire mountain rises with steep slopes, which also display another inside crater that is newer, and on very rare occasions, the center will display a red glow, indicating the presence of lava. So could it have been volcanoes in previous ages that erupted and dispersed the rocks that we found? I cannot think of any other means of their creation.

We deliberately and out of curiosity ran along such volcanoes, along their edges, and gazed inside their craters. Twice did we see radiating and flowing waves of lava.

Off to the side on one occasion we noted at the summit of a mountain an immense and high column of light consisting, apparently, of a large number of rocks that were burning so hot that they were radiating light as they ascended. The shock from their fall reached even to our feet at a distance.

As a result of the lack of oxygen on the moon or the result of other reasons, what fell to our feet were non-oxygen metals and minerals, and most of it was aluminum.

The lower and level plains, the dry seas in some places, were—contrary to the convictions of the physicist—covered with obvious but minor traces of Neptunian activity. We also enjoyed some of the dust that we caused with our feet walking on these lower plains. But we ran so fast that the dust remained behind us and immediately settled, since there was no wind that was actually lifting it but just our feet kicking it, and it did not get in our eyes and nose. We enjoyed it also when we set our feet on sharp rocky places and they seemed to us to be soft rugs or grass. None of the ridges posed any obstacle to our running in this region as they were no more than a foot or so high.

The physicist pointed with his hand at something in the distance and I saw on the right side what looked like a kettle discharging red sparks in all directions. From one aspect I could describe them as beautiful rays. So we decided to shift from our course to delve into the reason for this display.

When we arrived at the spot, we saw scattered pieces of more or less red hot metal. The smaller pieces were already cool, the larger were still hot.

"This is meteorite metal," said the physicist, taking one of the cooled pieces of the meteor with his hands. "Such pieces are what fall to Earth," he continued. "I have seen them in museums more than once. With no atmosphere on the moon, they stay hot for a long time, and are also unable to disintegrate. On Earth they can be seen when they enter the atmosphere and so become hot and disintegrate due to friction with the atmosphere."

Crossing the equator [toward the south] we again decided to move and now head for the north pole. The cliffs and piles of rock were impressive. Nothing like this existed on Earth.

We hurried and hurried, getting closer and closer to the [north] pole. The temperature in the ravines was decreasing. On the surface we did not feel this, because we steadily kept up with the progression of the sun. Soon we would have the opportunity to see its marvelous ascent in the west. We did not run too face, since there was no need to.

We did not go down into any ravine to sleep, because of the cold there, but rested and ate where we stopped. We even slept while we were in transit, abandoning ourselves to unrestrained dreams. This should not be a surprise, knowing that even on Earth such scenarios occur. It is possible all the more here where standing is the same as lying, due to the lesser gravity.

The Earth was descending, shining on us and on the lunar landscape more weakly. Then we reached the region where the Earth touched the horizon and started to slip behind it. This meant that we had finally reached the opposite hemisphere, the side not seen from Earth. What was uncomfortable was the thought of its lifelessness, motionlessness, its absence of growth. What a black and bleak background.

After about four hours, the Earth was hidden and we saw only a few mountain summits reflecting some light, and they finally became dark. The darkness was overwhelming. The number of stars was unfathomable! Only with a high quality telescope can this many be seen from Earth.

Then in the distance something was shining. After about half an hour we recognized this to be the peaks of mountains. So we ran to the top of a mountain and its other half was bright. There was the sun! But by the time we got down, it was already submerged into darkness and the sun was no longer visible. Obviously, this was the place of sunset.

We started to run faster, flying like an arrow released from its bow. But we did not really have to hurry; we would still see the sun rising in the west as long as we ran at a speed of about four miles an hour where we were. This was not running, so all we did was walk. No reason to hurry. And now was the marvel! The sun started to shine from the west. Its size quickly increased, and we saw a section of the sun and then the entire sun. It was rising, separating from the horizon. Higher and higher. Yet this was only occurring for us as we were traveling west.

"This is enough. I'm tired," jokingly cried the physicist, turning toward the sun. "Let's take a nap." We sat and waited until the sun descended in its normal way and disappear from view. "The comedy has ended." We rolled over and fell into a sound sleep.

When we awoke, again, but not hurrying, and solely for the sake of warmth and light, we chased the sun and did not allow it to disappear from view. It would rise and set depending on our speed, but it still kept us warm. We drew closer to the pole.

The sun being so low caused such immense shadows, and trying to outrun them caused us to get cold. In general the contrast in temperature was striking. We encountered one exceptional spot that was so hot we could not get very close. In other places, it was only one side of the rocks, the side facing the sun, that would get hot, and the other side was still cold, something that never happens in the polar regions of Earth.

So moving in the direction of the sun and making a circle, or more correctly, a spiral toward the north pole of the moon, we decided to depart from these frozen regions scattered with overheated rocks with barely a few comfortable spots.[1]

Many more hours passed. Although these places and mountains are never seen from Earth (being on the far side of the moon), they attracted our curiosity and then seemed just uniform and uninteresting anyway. The view was beautiful, looking at the distant Earth. We really needed to return to our residence and we now had no time for sleep. We wondered what was awaiting us at our camp.

So why now this melancholy feeling, this lonesomeness? Earlier we did not feel this way. Looks like our interest in the lunar surroundings had subsided, the interest in such a novelty was saturated. Quickly we turned toward our residence, ignoring the stars and sky from the moon's backside.

I felt our residence was not far off. But for some reason nothing was familiar to us, we could not locate any landmarks from our original departure. We could not find our yard anywhere and we could not recognize one view or mountain that was supposedly near our residence.

We walked and searched. Here and there. It was nowhere to be found. In despair we sat down and immediately fell asleep from exhaustion. The cold then woke us. We fortified ourselves with some food, with whatever was left, and continued traveling. With the sun having set and passed us, we overcame the cold by running, and for some reason, we did not encounter

[1] After crossing the moon's north pole the travelers now start southward, but on the opposite side of the moon, toward their camp.

any ravines where we could shelter ourselves, where it was warmer than the surface. So again we started to run after the sun, but the sun seemed so far away and we could not catch up with it.

We looked in our bags and there was only enough food left for one more meal. What to do, then? We ate what was left, our last meal. Want of sleep was forcing our eyes to close. The cold compelled us to huddle closer to each other. So what happened to all those ravines that we had encountered earlier, now that we needed them? We only slept for a short while: the cold was worse and caused us to wake up. We hardly slept three hours and it was not enough.

Weak, exhausted by worry, hunger and the overwhelming cold, we could not travel as fast as we did earlier. We froze. We held each other tightly, not to let the freezing cold cause our death. Still we started to slumber and huddled closer to one another. The physicist fell asleep and dreamed of Earth. I held him closer, to not lose heat.

Deceiving dreams of a warm bed, a burning fireplace, hot food and a bottle of wine took control of me. My family was surrounding me, they were walking with me.

*

Dreams and day-dreaming! Blue sky, snow on the neighbors' roofs. Birds flying. People, faces of friends. The doctor. What is he saying?[1]

"Loss of consciousness, a prolonged faint, a serious condition. Significant loss of weight. Exhaustion. But nothing to worry about now. His breathing is better. His sensations are returning. The danger has passed."

Surrounding me I hear joy, although the faces are filled with tears.

To say this in brief: I had been unconscious and now awoke. I had laid down on Earth and awoke on Earth. My body had remained here while my thoughts were on the moon.

Nonetheless I remained in a delirium for a long while. I asked about the physicist. I talked about the moon. Surprisingly, I even asked how all my friends got on the moon. I had the earthly interwoven and confused with the lunar. I imagined myself on Earth, but was still returning to the moon.

The doctor told them not to interfere with my statements or agitate me. Slowly I returned to consciousness and gradually recovered.

The physicist was standing there with my other friends. After I regained my health, I narrated to him my adventure and he was surprised. He advised me to record it and supplement it with some explanations.

[1] This is where our subject awakes from his dream about being on the moon.

Appendices

He Lived Among the Stars: A Eulogy on behalf of Konstantin Tsiolkovsky

By Aleksandr Romanovich Belyaev

Today completes five years from the day of the death of Konstantin Eduardovich Tsiolkovsky. The country has not forgotten and will not forget the bold inventor and self-taught scholar. This occasion of five years from the date of his death is dedicated to the opening in Moscow of this memorial on his behalf. His compositions have been published, the complete set in six volumes, and are widely purchased by members of the most diverse spheres of science and technology, even those which seem less significant, dealing with bacteria, how to make use of ocean tides, and others.

This calendar year we must remember likewise his faithful companion and wife, Varvara Evgrafovna Tsiolkovskaya, who barely outlived him by five years. She died this year on August 23. Her death forces us to remember the private and family life of the Tsiolkovskys, which is very informative. Much depended on his personal life, including so much of his scientific work. His family home was also his study, office and laboratory and workshop.

A certain portion of our young people, even though they might be considered a minority, after completing higher educational institutions, involuntarily leave the centers of their relatively provincial cities to work elsewhere, stating that work in their home towns was difficult and it would be impossible to find scientific work and grow there.

The lives of Michurin[1] and Tsiolkovsky testify that such reasoning and conclusion is completely unfounded. Both Michurin and Tsiolkovsky were self-taught, working in the most obscure and primitive corners of pre-revolutionary Russia, and nonetheless they advanced to the ranks of scholars having a world-wide recognition.

How were they able to achieve this? It was due to their love of science, their complete devotion and aspiration, a strong work ethic, the ability to ascertain every objective cause.

A few years back, at my request, the late Varvara Evgrafovna recorded her recollections of her life together with K.E. Tsiolkovsky. It was precise, yet written at an elementary level; even though she finished secondary education, she had little opportunity to write. She was concerned with domestic matters and daily chores, while at the same time her husband lived in outer space among the stars.

This is how she described the way the young Konstantin came into her life:

> Upon his arrival in Borovsk, Konstantin Eduardovich rented a room from us. He wanted to live fairly close to the river. He rented a large room, living room and side room from us. He did not place his bed in the side room, but in the large room, and this surprised as well as irritated me. He was concerned, wanting more fresh air. In the large room hung a model of an aerostat that was 12 feet long and rather narrow, made of writing paper. He wore a long coat made of some inexpensive cloth, a flannel hat with earmuffs and a red scarf. His hair was long. He wore earmuffs because his hearing was poor and he needed to protect his ears from the cold. At the time, no one wore earmuffs and as a result people would turn their attention to him.
>
> Then he proposed marriage to me, and on the morning of August 20, we married. We had a modest wedding, no one accompanied us as would be proper. K.E. did not like any formality or ceremony. That afternoon he went to school anyway and surprised the teachers and students. They told him, "At least this one day you can skip teaching in school." Later that same day, a turning lathe arrived at our house. In one respect the entire situation that day was rather humorous; he was in a hurry to reach the region of the stars.

Family life was distracting to E.K. some of the time, as he was in a hurry to ascend to the special regions. When he made his marriage proposal to Varvara, he warned her that their life would not be easy, and they would have few friends who would visit them, and they would not go about visiting others. It would be a modest and hard-working life. And so it was.

---

[1] Ivan Vladimirovich Michurin (1855–1935).

Konstantin's aunt wrote that they would expect him in Ryazan when he was on vacation, and especially his sister Masha, but he never arrived. Konstantin Eduardovich was busy with his experiments. He ordered from Moscow various retort bottles, glass tubes, a telescope, a microscope, thermometers, barometers, and soldering equipment, in order to make a steam engine.

A significant portion of his income he squandered on chemistry equipment and instruments, and to publish his books, which he would distribute without charge. The entire arrangement of their family was such that it would support the continuation and success of his scientific experiments. He lived in the sky more than he lived on the ground. This was a genuine citizen of the universe. One could scarcely find such a person among astronomers who so revered outer space as did E.K. Tsiolkovsky. He accomplished his imaginary journey through his science fiction story *Beyond Earth*, where he clearly displayed how gravitational force varies on various planets and asteroids. His entire life he struggled with a means of overcoming Earth's gravitational force. Airplanes, aerostats, stratostats, and rockets were for him only stages in the aspiration of a person to reach the stars. He likewise made his entire family familiar with the regions of outer space.

Konstantin Eduardovich regularly spoke to his wife Varvara Evgrafovna about the stars, the moon; and it was interesting to her, a novelty, and she suggested he publish his efforts, since others might also find this interesting. He would talk about his experiments during dinner or at tea.

His daughter, Lubov Konstantinovna, who was also his secretary for many years and his assistant, records the following in her memoirs.

> Father loved nature and especially the sky. The science of the universe he considered fundamental to all branches of knowledge. On star-filled nights he would take his telescope and gaze into the heavens. My brother and I, and sometimes some of the neighbors' children, would be part of this star-gazing. We would ask him questions and he was more than happy to answer all of them.

It was not easy for him to live under the regime of imperial Russia and it was difficult likewise for his family. The amount of his income that he dedicated to his experiments were a sacrifice he made on behalf of heaven. In later years the Soviet government provided monetarily for K.E. Tsiolkovsky so that he could continue his experiments, and after his death, the government provided his wife and his daughter a considerable personal pension.

Hard times pass, but great feats remain.

K. E. Tsiolkovsky's Major Compositions

Tsiolkovsky composed over 400 articles, stories, books, and pamphlets over his lifetime. About 180 of these were philosophic, about 20 were science fiction, and the remaining 200 were technical and scientific. The following are the most popular or most important.

A Navigable Metal-Clad Aerostat (1891)
A Traveler into the Spatial Expanses (1938)
About the Soul, the Spirit and the Cause (1923)
Adventures of the Atom (1917–1918)
Assurance of Life, The (1934)
Basic Physical Hypotheses (1933)
Beyond Earth (1896, 1918)
Calm and Joy (1920)
Catastrophes of the Earth (1921)
Cause of the Cosmos (1935)
Change of Gravity Relative to Earth, The (1894)
Citizen of the Universe (1933)
Conditional Truth (1932)
Cosmic Journeys (1933)
Cosmic Philosophy (1935)
Cosmos is Alive, The (1932)
Debates over the Cause of the Cosmos (1923)
Development and Restoration of the Universe, Cycles of the Universe (1918)
Discussion of Human Rights on Earth (1933)
Domination of Life and Intellect, The (1932)
Doubtfulness of Every Philosophy, The (1934)
Duration of the Sun's Radiance, The (1897)
Earth's Energies (undated)
Earth's Energy, The (1932)
Earth's Ethics, The (1934)
Eternal Activity of the Universe, The (1933)
Ethereal Island, The (1928–1930)
Ethics (1903)
Exploration of Outer Space Using Jet-Propelled Rockets (1903, 1911, 1926)
Fate, Destiny, Fatalism (1919)
Formation of the Solar Systems (1929)
Freedom of the Will (1918)
Future Earth and Humanity, The (1928)
Future Earth, The (1928)

Genius among the People (1918, 1921)
Goals of Inter-Stellar Travel (1929)
Graphic Portrayal of Feelings (1880)
Gravity as the Primary Source of all Universal Energy (1893)
Guides of Humanity (1929)
Higher Truth (1932)
How Long Has the Universe Existed? (1920)
Ideal Mode of Life, The (1917)
Initial Cause, The (1918)
Intellect of the Cosmos and the Intellect of Its Entities, The (1933)
Is a Metallic Aerostat even Possible? (1883)
Is there God? (1932)
Is there Spirit? (1932)
Joy and Suffering (1924)
Joy without Expense (1923)
Kinetic Theory of Light
Life in Inter-Stellar Space (1933)
Life in the Cosmic Ether (1924)
Life of Humanity (1930)
Living Cosmos, The (1939)
Living Entities in the Cosmos (1895)
Living Universe, The (1923)
Longevity (1934)
Love of Yourself or True Ambition (1928)
Mechanics of Peace, The (1923)
Mind and Passion (not dated)
Mirages of the Future Societal Organization (1918)
Monism of the Universe (not dated)
Movers of Progress (1927)
My Ideas of Monism (1924)
My Philosophy (1932)
Necessity of a Cosmic Point of View, The (1934)
Nirvana (1914)
Of the Possibility of the Construction of a Metallic Aerostat (1890)
On Alien Planets (1905)
On the Moon (1883)
On Vesta[1] (1930)
Order of Cosmic Philosophy and Its Conclusions, The (1933)
Organic Formation of the Universe, The (1932)
Organization of Society, The (1918)

[1] Also known as 4 Vesta, an asteroid designated a minor planet.

Perfection of the Laws of the Universe, The (1928)
Pervious Earth, The (1928)
Plain Meditations on the Eternity of Matter and Sensation (1933)
Planets Are Populated with Living Entities, The (1933)
Qualities of the Cosmos (1932)
Regarding Monism, Questions and Answers (1920)
Regulations of Society (1919)
Rights and Obligations of Humanity (1933)
Science and Belief (1917)
Scientific Basis of Religion, The (1898)
Scientific Ethic (1927–1928)
Second Law of Thermodynamics, The (1914)
Self–Generation (1929)
Sensation, Life and Death (1934)
Significance of Industry, The (1934)
Sketches of My Life (1935)
Socialist Organization (1931–1932)
Sorrow and Genius (1916)
Stages of Humanity of the Transformation of Earth (1927)
Subjective Uninterruption of Higher Forms of Life, The (1933)
Technical Progress (1932)
Theorems of Life (1928–1930)
Theories of the Cosmic Eras (undated)
Theory and Experiments of an Aerostat Operating in a Horizontal Direction with an Elongated Shape (1885–1886)
Theory of Aerostats (about 1887)
Theory of Gases (1881)
Thoughts on a Better Societal Structure (1903, 1911)
Uninterruption of Life, The (1933)
Unknown Intellectual Forces (1902–1903)
Unrestrained Spatial Regions, The (1883)
Value of People, The (1934)
Wealth of the Universe (1919–1920)
What Type of Government do I Consider the Best (1934)
When Will the Sun Extinguish? (1933)
Will of the Universe (1928)

Bibliography

Арлазаров, Михаил, *Циолковский,* Приокское Книжное Издательство, Тула, 1977.

Циолковский, Константин Эдуардович, *Грезы о Земле и Небе,* Приокское Книжное Издательство, Тула, 1986.

Циолковский, Константин Эдуардович, *Моя Жизнь и Работа, Цели Звездоплавания,* Либроком, Москва, 2011.

Циолковский, Константин Эдуардович, *Философия Космический Эпохи.* Академический Проект, Москва, 2014.

Циолковский, Константин Эдуардович, *Миражи Будущего Общественного Устройства,* Луч, Москва, 2011.

Printed in the United States
By Bookmasters